EPISODES

Salim Akhtar

CENTRAL PARK SOUTH PUBLISHING

Published by Central Park South Publishing 2023
www.centralparksouthpublishing.com

Co-editor: Ehtesham Haque
Typesetting and e-book formatting services: Victor Marcos

About the Author

After finishing college, Salim Akhtar joined the Pakistan Army, serving there for a short tenure when he witnessed the 'Nagi Battle' between Indian Army and Pakistan Army. The eastern wing of Pakistan separated in 1971 to become an independent country, Bangladesh. Salim moved on to join the Bangladesh Army, where he served for thirty-plus years. His military experience extends across conventional war, counter-insurgency operations and United Nations (UN) Peace Keeping operations. In the military, he spent significant time on teaching assignments, also being the Commandant of Bangladesh Military Academy. Salim was the commander of the Bangladesh UN Peacekeepers contingent of 1,220 men in Bosnia Herzegovina. He was in Bihac, Bosnia Herzegovina, during the height of the conflict from September 1994 to November 1995. He participated in the implementation of the Dayton Peace Accord. During the war, Serbs were in control of the access roads into Bihac, so the international media did not have access into Bihac. Salim was frequently sought-after person by the media at the time for interviews or for queries related to ongoing war events inside the Bihac enclave which was a UN-declared safe area. Salim carries with him a trove of information on the events during the war, and, more importantly, information on the political dynamics which influenced the UN Peacekeeping Operations. In 1995, his interviews were published in newspapers like The LA Times and on television channels like BBC, Sky TV UK, UN TV, Bangladesh TV and on 'Peter Jennings's show' ABC TV USA.

Colonel Salim Akhtar
Commander Bangladesh Contingent
UNPROFOR, Bosnia Sep 1994—Nov 1995
Retired as Brigadier General Salim Akhtar

Contents

About the Author. iii

Preface vii

Chapter 1. 1
- *Childhood Years*

Chapter 2. 7
- *Service with Pakistan Army*

Chapter 3. 21
- *Nagi Post Battle Western Front 1971*
- *36 Frontier Force Regiment Pakistan Army—4 Para Battalion Indian Army*

Chapter 4. 37
- *Prison Camps & Repatriation to Bangladesh*

Chapter 5. 45
- *Bangladesh Army—Turmoil in Formative Years*

Chapter 6. 49
- *Service with Bangladesh Army*

Chapter 7. 125
- *UN Peace Keeping in Bihac Bosnia Herzegovina*

Afterthoughts. 245

Index/Acknowledgements. 269

Preface

I have been planning to write this book for many years now. My initial manuscript covered only the wartime events of Bihac, Bosnia Herzegovina. I was motivated to write the book because I felt that much of what happened in the conflict zone in Bihac, Bosnia, had never been covered in the media due to access restrictions. Bihac, located on the northwest tip of Bosnia Herzegovina, was an isolated Muslim enclave surrounded by Serb-held territory at the time. Even the UN convoys required prior travel approval from the Serbs for entry or exit. During my stay of fourteen months, I encountered visits by only two media teams. One was a UN TV Team, and the second was a single reporter from Sky TV UK who managed to sneak into the area on a Bosnian night infiltration helicopter flight. It was a hazardous venture, with the helicopter dodging the Serb's air defenses. One such night-helicopter did get shot down during the war, and all on board were killed. The only source of information for the international media was either the news briefs released by UN sources or the Serb media. Given the media's inadequate news content, I decided to hold onto my records to publish as a book at some later time. Unfortunately, it has taken nearly 25 years due to my own indecision. Finally, in response to a great deal of prodding from my friends and colleagues, I have written my record of events, especially as pertains to the Bangladesh Army.

In this book, I have used my notes, my memory, my wartime diary, and some official documents, leaving out the source identification so as not to attribute blame or responsibility. I know many books have been written covering the Bosnian war, but to my knowledge, the inside story of Bihac is yet to be covered. In that sense, this may be the first book from somebody personally involved in events occurring during the war. Later, when I finished my manuscript, I accepted the suggestion to convert it into a kind of story of my life, starting from childhood to the making of a soldier. I found it interesting to cover the events of my career with the Pakistan Army and later with the Bangladesh Army. After listening to suggestions from my friends, I redrafted the entire manuscript into my memoir, with the Bosnian war being a significant part of my life story. I have also covered my short career with the Pakistan Army in reasonable detail, as that is where I was trained and commissioned as an officer. I have added significant details about the ' Nagi Battle ' fought on the western front after the ceasefire between the Pakistan Army and the Indian Army as I witnessed this fighting from frontline trenches. I am happy about the book finally evolving as my own memoir. The opinions expressed in this book are solely my personal views and should not be attributed to any country, organization, or individual. They do not represent official views or policies at the time. The inclusion of wartime photographs and documents in this book is based on materials from my personal records.

I remain grateful to the officers and soldiers of the Bangladesh Battalion (10 East Bengal Regiment) who displayed immense courage and resilience during the harsh Yugoslav winter. Despite facing shortages of food and fuel caused by the Serbian denial of UN logistics convoys into the area, they never wavered. They endured the hardships of being separated from their families, receiving only occasional letters from home, yet they never complained. The Bangladesh soldiers demonstrated extraordinary resilience and unwavering determination, even in the face of repeated hostile fire from the Serbs. They steadfastly held their ground, defending their positions. My thoughts and prayers are with the brave Bangladesh soldiers who lost their lives or were injured due to targeted attacks by the Serbs on UN personnel. Their sacrifice will always be remembered.

I am grateful to my wife for taking care of our family with minimal support and enduring the fourteen months of my absence. I am also thankful to my friends and former military colleagues for assisting me and encouraging me to write this book. Finally, I am grateful to my eldest son Ehtesham for painstakingly editing the book and for providing the much needed support until the book was published.

Colonel Salim Akhtar
Commander Bangladesh Contingent
UNPROFOR, Bosnia Sep 1994—Nov 1995
Retired in 2004 as Brigadier General

CHAPTER 1

Childhood Years

As a young boy, I never dreamed of becoming a foot-slogging infantry soldier. However, I trust my destiny took me on the path of soldiering instead of that of my childhood dreams. My life goals kept changing, yet I adapted to what was practically feasible and successfully pursued that end goal. At the end, I believe I was successful in my judgement, because my work was valued and appreciated by the organizations for which I worked.

My parents never thought of me becoming a soldier, which to them meant putting one's life in harm's way. My father grew up in his village in the most modest environment, at a time when Bengal was part of British colonial India. However, Japanese military advances through Burma during World War II resulted in enhanced military activities in Bengal. My parents came from a small village on River Halda at Katirhat, Hathazari, Chittagong. After finishing school, my father decided to join the Royal Indian Air Force in 1944. Incidentally, the Royal Indian Air Force at the time had an airstrip near our village at Hathazari, Chittagong, providing support to the British Army operations in Burma. And that may have attracted my father to join the air force.

I remember asking him why he did not opt to be a fighter pilot, which was a job that was in great demand during World War II. He explained that having grown up in the village, all he knew about flying was the scary sound of Japanese aircraft on bombing runs. He was afraid of airplanes, and pursuing warfighting would not have entered his mind. He wanted a peaceful desk job, well away from the war front, and that's why he got the job in the air force accounting branch. He was trained and posted to the Royal Indian Air Force base at Barrackpore, near Kolkata in West Bengal, India. A few months later, he was transferred to the Royal Indian air force base at Ambala, a city in northwest India in Punjab. In 1947, after the partition of India, being a Muslim, he opted for the newly independent country of Pakistan. His subsequent journey from Ambala in India to Pakistan is a tale of the mass refugee migration. My father survived the journey on those infamous 1947 migration trains, which transported millions of refugees across the Punjab border into Pakistan. That was his first migration, travelling further away from home and moving into Pakistan to continue his job with the air force. Many years later, after the independence of Bangladesh in 1971, he undertook his second migration with us, the family members. This time he was returning to the homeland he had left in 1944. He was returning to the newly independent country of Bangladesh on special repatriation flights organized by the ICRC (International Commission of Red Cross).

Post 1971 independence war, my parents, along with other Bengali air force families, were interned for two years at the Warsak repatriation camp, which housed family members of Bengali air force officers. After two years in the camp, my parents were repatriated to the new nation of Bangladesh. My father joined the Bangladesh Air Force in January 1974, the third air force he was to serve. He retired at the Rank of Squadron Leader from the Account Branch in 1979.

I was born on March 31st, 1951, in Kohat, Pakistan, where my father was posted. Kohat is a city located in the northwest frontier province of Pakistan, inhabited by Pathans. I do not remember anything from my early childhood days spent in Kohat. My recollections begin from the days when I started going to school, probably at the age of four. Around the mid-fifties, my father served at Mauripur Air Force base in Karachi, later renamed as Masroor Air base. The base was renamed Masroor in honor of a B-57 bomber pilot who died in an air crash at the base caused by a bird-hit. The bird struck the plane and went through the canopy, hitting the pilot right on his helmet-protected head. May God bless the soul of the pilot and the killer bird. As a young boy, I often visited the air base vicinity to watch aircraft take off and land. The aircraft we saw were World War II vintage; however, of interest to me were the troop-carrying glider flights towed by other aircraft. Many years later, I read a great deal about the use of these troop-carrying gliders used for troop insertions during the second World War. Watching the planes fly was exciting for my young friends and me, making me dream of becoming a pilot, something I never could make happen.

Since my father served with the Pakistan Air Force, my schooling was in Pakistan. I was possibly only four years old when my parents sent me to school straight into class 1. I started schooling in St. Patrick's School Mauripur junior section, the senior section was located a distance away. I remember the first day my father left me at school. I ran out of my classroom crying and running after my father trying to catch him. That was day one; later, I spent four good years at that school. It was a Catholic school, so before the start of classes every day, we prayed, seeking blessings from God. Prayers were led by our female class teachers or Sisters, as we called them. My school was one hour walk away from home across a sparsely populated area. So, every day with a few friends, we walked up to the school, playing and creating all kinds of mischief along the way. I remember that we would put iron nails on the railway line while going to school in the morning and check in the afternoon as we returned home to find the nail flattened by the rail wagons having gone over them. On the way to school, my friends and I would look for water holes to catch tadpoles and small fish, which we would put in a jar; they probably survived a few days.

An incident that still makes me laugh was the payment of my school tuition fees. My monthly tuition fee for Class IV at St Patrick's School was nine Rupees, so my mother gave me a single note or bill of ten Rupees to pay the tuition fee. I was permitted to keep the change for buying tiffin snacks. Instead of paying the tuition fees first, I went to the school canteen and bought something to eat. I got into trouble when the canteen guy returned me a massive load of change in coins. With that load of coins, I sheepishly went and paid the school fees. I cannot remember if the person at the school accounts office scolded me or not.

Being the child of a military person, I had to endure the difficulties of frequent school changes. Every three or four years, my father was transferred to a different location, and I was forced to change schools. Later, while serving in the military, my sons faced similar difficulties. I remember finishing Class IX from a school in Karachi in 1964 when my father transferred to Rawalpindi. I was admitted again in Class IX in 1964 at Sir Syed Public School Chaklala Road, Rawalpindi. I repeated Class IX because Karachi School Board and Sargodha School Board had different syllabi and exam systems. So, I wasted one school year. A year later, after the 1965 war, my dad was posted back to Karachi but luckily, this time, I did not have to lose an academic year as I had attended Class IX under Karachi Board earlier. I finally passed my SSC (Secondary School Certificate—Class X) from Cantonment Public School, Malir Cantonment Karachi, in 1966. A year later, in 1967, we moved again; this time, my father was transferred to PAF Public School Sargodha to work as a bursar. The very next year, he was transferred again to nearby Sargodha Air Force Base. Sargodha, at the time, was the largest air base for the Pakistan Air Force with the American F-104 Star Fighters and the freshly inducted squadron of French Mirage III fighter bombers. My friends and I always took advantage of the opportunity to go and see the aircraft's firepower demonstration on a nearby firing range, and we never missed going to the air power demonstration shows.

I studied at Government College Sargodha and finished college in 1968. I was the only student in the class from an English school background; the rest all hardly spoke English and dressed up in kind of rural Punjabi dress. The other predicament for me was that I was very weak in Urdu because I studied 'easy Urdu,' which was the practice in English medium schools. All students from Urdu schools at the college studied regular Urdu literature. I found it very difficult to cope with Urdu literature being taught at the college. The two years at the college were not so enjoyable for me to remember anything. I simply attended classes and returned home because I hardly made any friends in college. I had a few friends in the locality where we lived, and with these friends, I mixed well. I finally passed my HSC (High School Certificate—Class 12) from Sargodha Government College in 1968. I enjoyed studying chemistry and, at the time, did give serious thought to studying chemical engineering, but there was no university nearby. Going to university would have required me to relocate to the city of Lahore and to stay in dorms. All that extra expense would have been burdensome for my father, with six children to look after at the time. He also supported his extended family in his home village. So, the idea of going to university quickly vanished, and I decided to apply for the air force to be a fighter pilot, which was my childhood dream. Unfortunately, I never made it to the air force because I was twice declared medically unfit for being grossly underweight. I decided to move on and take an alternate career path.

CHAPTER 2

Service with Pakistan Army

After my selection in the army, I joined the Pakistan Military Academy at Kakul in November 1969 with the 45th PMA Long Course. To join the academy, I travelled overnight from Sargodha to Rawalpindi, arriving in the morning. I knew from my friends who had joined the army earlier that if I had taken the morning bus to reach the academy by lunchtime, I would spend the rest of the afternoon and late into the evening being ragged. Ragging is a traditional practice of harassment and punishment by senior term cadets on junior cadets. In the military academy, senior cadets are like sharks in waiting for the new cadets to arrive. I was smart; I left my luggage in the locker room at the Rawalpindi railway station and spent time till mid-day window-shopping around the city's Saddar shopping area. Later, I took the afternoon bus to reach the military academy in the evening, thus effectively cutting out the extra ragging time from my seniors, who I knew would be waiting like hounds ready to pounce on new arrivals. So, on the first day, I did manage to escape much of the ragging by senior cadets.

Training at the military academy went on for two long years. It was physically and mentally demanding. What we feared most was the after-hours ragging by the senior cadets. During the two years at the academy, we had four training terms, each of about six months with four weeks of inter-term break. A training term is more like a semester at a university. In the academy, the cadets are organized as follows. A new batch or class has a numeral ID. For example, my batch is named 45th PMA Long Course. Each course can have 130—150 cadets or more. Cadets are organized into smaller groups called platoons, like a class group. Each platoon will generally have around twenty-five cadets. So, the entire batch or course comprises of 7—8 platoon groups. Each platoon is placed under the supervision of a Term-4 or final term senior cadet who holds the appointment of Platoon Sergeant. The Platoon Sergeant is assisted by three Term-3 cadets holding the position of Platoon Corporal. Together these four senior cadets take charge of the platoon of twenty-five new cadets, supposedly to groom up the new cadets.

During our first term, we had an Iraqi Corporal, Nezar, who possibly died during the Iraq war in the 80's. This guy was tough and loved to take us for morning runs well before daybreak and before the formal academy training started. On weekends Corporal Nezar enjoyed the sadistic trick of not letting us have sleep or rest. He would get the whole lot of us, Khaled Company Platoon K4, standing in front of his room and make us do all sorts of physical punishment. This included everything from frog jumps, repeated push-ups, running around the barracks, and changing dress.

The practice of changing attire or uniforms in quick succession can be very punishing mentally. The new trainees are told to first appear, maybe, in sports attire; once they do that, the whole group is told to go and change into some other attire which could be, say, a battle uniform or a dinner attire. Then trainees are required to run up to room, dress up in the new form and report back in five minutes, indeed a difficult task. So, you run up to your room, take off your clothes as fast as you can, throw the clothes all over the floor, put on the specified dress and run back to the reporting place only to be told to go back and change into a formal suit or some other form of attire and report back again in five minutes. Not all cadets can make it in the given five minutes. Latecomers end up getting nonstop punishments. While at the academy, changing attire seemed like a punishment, but honestly, this proved to be very helpful later in life as we got used to dressing up in a very short time for short-notice events.

At the academy, I was in the Khaled Company accommodation barrack K-12. These were World War II period wooden barrack facilities for the soldiers of the old British Indian Army. The accommodations had no heating, no running water and a bucket toilet. The toilet had a back door for the water man to fill the metal water tub and clean the toilet every morning. During the winter, the water in the metal tub would be ice cold, so I never washed my face in the mornings. I used to just wet my fingertips to wipe my face. Often to escape the weekend group punishments, I would lock my room from the outside, enter the room from the rear toilet cleaner's door, and quietly sleep in my bed. I can still remember being reported as missing and with the room locked from the outside and others inquiring the whereabouts of Salim. I guess I was a rebel and not a rule-abiding boy at the time, frequently doing forbidden things. Having stayed the first year in wooden barracks, we later moved to newly completed accommodation, a multi-storied building with heating, hot water and other amenities. The downside of staying in the new dorms was that we were now staying in close proximity of the senior cadets. The moment you came out of your room, you would encounter senior cadets in the corridors which would result in more frequent punishments.

I can recall a few reckless actions that I took during my training at the military academy, and as I got away with them, I grew bolder and began to undertake even more daring ventures, known as "pangas" in the local military slang. One such incident occurred during my first term when we were allowed to watch movies in the academy on weekends. The cinema hall had designated seating for cadets from different terms, with the first row reserved for first-term cadets and subsequent rows for the higher terms. As the seating for senior cadets began to run out, some whispered to the junior cadets to vacate their seats for the seniors. This process continued, with the front row of first-term cadets being forced to leave.

Although I lost my seat, I was determined to watch the movie. I came up with a creative idea and walked to a nearby hockey field and waited for the movie to start. When the movie began, I slowly walked up to the rear side of the building and knocked at the projection room door requesting the crew to let me in. I told them I wanted to see the movie through one of the spare small projection room glass apertures sitting on a wooden tool. I wonder whether the projection room crews were shocked or surprised or took me for a ghost. The projection crew was friendly and allowed me in but cautioned me not to make any noise or get too close to the glass window as the academy commandant was sitting below it. They also warned me when five minutes were left before the interval and the final end of the movie. I quietly left before the interval and watched the crowd from a distance, waiting for the movie to restart. After the interval, I re-entered the projection room and watched the rest of the movie leaving before the end, disappearing into the darkness towards my dormitory. Looking back, I realize that I always had a strange attitude of wanting to do forbidden things and take on challenges.

Here is another interesting incident I remember from my third-term days. We were out in the field for three or four days for military tactical exercises. On day 1, after a day long march and clearing simulated enemy oppositions, we arrived in the evening at a location where we were told to take up defensive positions. I was a simple rifleman in a forward squad position in the exercise setup. Like all the other cadets, I had the task of digging my fire trench through the night, which would be inspected the next morning. The requirement was to dig in before the simulated attack expected the following day. That night there was a slight drizzle, and it was a wet and cold winter. Being out in the open field and digging the trench was a tough job on a wet winter night. I, therefore, decided to find a dry and warm place for a comfortable night's sleep. I walked up to the village in the front, looking for shelter. Without disturbing the villagers, I found a cattle shed with a donkey inside staring at me. This cattle shed was a half-dug mud-hut, which was warm inside. I decided to stay the night in this cattle shed and slept in a corner over a haystack. I slept through the night but made sure to wake up well before daylight and be present in my battle position to dig a little bit to show I had been doing the job assigned to me through the night. The day before had been tough walking all through the day, so everyone slept the night quietly, and nobody noted my absence.

My thirst for adventure did not end there, as I continued to seek out new exploits growing more confident. However, one particularly mischievous act during my third term had severe consequences for my military career. This single incident caused me to lose my seniority, which is a significant issue in the military. It happened the day before Eid holiday, and my platoon mate Farooq Afzal and I were eager to leave the academy early to catch a train to Rawalpindi and reach our respective homes before the Eid Day. We decided to skip the last two classes and asked the senior platoon cadet, Matin, to report us as sick and present at the academy hospital Medical Inspection Room. Our plan worked initially, and we changed into civilian clothes and sneaked out of the academy under the wire fence near the physical training gymnasium. However, our luck ran out as we were spotted by our platoon commanders, who were returning from Abbottabad city. They reported us missing, and we were punished by being relegated to the next junior class, the 46th PMA Long Course, losing one year of seniority in the process. This incident had far-reaching implications for my entire military career as my juniors became my seniors overnight.

In the 1960s, telephones were not easily accessible, and as a result, communication with my parents, who lived in a distant air force base in Sargodha, was rare after I joined the military academy. During the late 60s and 70s, telephone calls were transmitted through physical copper lines, with connections made through multiple exchanges in different cities. The current generation would find it difficult to fathom the process involved in making a long-distance call, which required calling the telephone exchange reservation number and booking the call. One would often have to wait for an hour or more for the telephone operator to call and request you to hold as they attempted to connect you to the other end. It was a harrowing experience to make a long-distance call through landlines at that time, involving multiple telephone exchanges, with the operator shouting as the other end repeatedly stated that they couldn't hear you. However, because there were no direct dial services or cell phones at the time, we had no option but to endure these long waits and frustrating calls.

Pakistan Military Academy Term 1. Saluting Test. Author third from right, Dec. 1969 (Source: Author's own records)

**Pakistan Military Academy Term 1. Bengali Cadets Platoon K4.
Author first from left – Dec. 1969 (Source: Author's own records)**

After six months of rigorous training at the military academy, I was in my second term and decided to join the parachuting club. Every cadet aimed to get selected for para jumps, but only the fittest and most determined were chosen. Luckily, I was one of the selected few for training at the Pakistan Army Para School in Peshawar. It was during the autumn semester break of 1970 when most cadets headed home for the long-awaited four weeks of vacation. However, a few of us crazy youngsters, including me, chose to undergo adventure training instead of going home to meet our parents.

The training at the Para School spanned over three weeks, during which we had the opportunity to make five jumps from C-130 aircraft using US Army T-10 parachutes. After the para training, we had a few days of the fourth week to visit our parents before returning to the academy. I remember those training days with great respect under the officer in-charge at the Para School, Major Tariq Mahmud, who was popularly known as TM. He was from the special forces and, years later, as a Brigadier General, he became the Head of the Special Forces group (SSG) of the Pakistan Army.

However, Brigadier General Tariq Mahmud's life ended in a tragic incident during a national day parade, in front of a massive crowd, including his wife and children, who watched him drop to his demise. He was possibly past 60 years old when he died doing a free fall jump. Unfortunately, both his main parachute and reserve parachute failed, and he dropped like a stone to the ground. Anyone who served or trained under the late Brigadier Tariq Mahmud will remember him as a very inspirational leader. Para school physical training was very tough. Often during the morning physical training class, Major TM would be the demonstrator himself in the front of the class doing the push-ups and other exercises with us. We learned during the training that one of the risks after a touchdown is parachute drag, caused by high wind conditions. If the parachute catches air, it becomes a sail and can drag the person over a long distance to death unless one learns to recover quickly. We were told that one such incident did indeed occur during jumps at the para school because the jumper failed to recover his parachute on landing and prevailing high winds dragged him to his death.

For quick recovery during high-wind landings, we practiced strapping on the parachute harness and lay on

our backs on the ground. Only a thin protective pad at the
back of the parachute harness protected the body from
touching the ground. With the harness strapped, we were
hooked to the rear bumper of the jeep. The jeep would
drag the trainee jumper till the jumper recovered and stood
up, running behind the jeep. The trainees who failed to do
the required recovery drill properly, ended with severely
bruised elbows, and Major TM would smile and comment,
the wind just got faster, implying he had driven the jeep a
bit faster. He was doing the right thing, training us for the
difficult and life-threatening high wind landing scenario.
We all felt afraid when Major TM was driving the jeep
because he would deliberately drive faster.

Fortunately, I never had a bruised elbow and always
recovered from the drag quickly. During my five jumps,
twice I was the jump group leader, called a stick leader which
required me to shuffle and stand next to the airplane's open
door for about thirty seconds before exiting. I must admit
standing at the airplane's open door as a jump stick leader
was scary because the airplane keeps on pitching and rolling
slightly and you get the feeling you will drop out. On green
light the jump master would shout 'go' and the jump would
start with each jumper exiting one after another with few
seconds gap, idea being remain as close as possible to land
in a close cluster. We were dropped from an altitude of 1200
feet, and I, being the lightest in the group, was one of the
last to touch down, taking about 55 seconds to descend. I
had noticed that with the open aircraft doors on both sides,
the onboard jump instructors always wore backpack chutes
as safety in case one slipped out, but Major Tariq Mahmud
never wore any safety parachutes. He would wear a white
half-sleeve tee shirt tucked into his military trousers. As

jump master, he would hold the open-door sides sticking his head out, taking the full blast of engines while trying to look out for landmarks to see if we were close to the dropping zone. Standing next to the door as a stick leader, I always wondered what if he fell out, and ironically, his death came from a free fall jump many years later. Throughout my military career, I proudly wore the parachute wing pinned on my chest by Major Tariq Mahmud, who lived to be a great legend. In honor of our instructor Major Tariq Mahmud, I have included the picture of my parachute jump logbook signed by him.

Finally, after two years of gruesome training, from November 1969 to September 1971, at the Pakistan Military Academy Kakul, I was commissioned as a Second Lieutenant (2 Lt) in September 1971 and was posted to 36 Frontier Force Regiment (36 FF) based in Quetta.

Pakistan Military Academy Term 2. Assault Course Training. Author on the left – Feb. 1970 (Source: Author's own records)

Pakistan Military Academy Term 2. Boxing. Author on the right –
Sept. 1970 (Source: Author's own records)

Pakistan Military Academy. Parachute Club Jump at Peshawar Para
School. Author on foreground – Oct. 1970 (Source: Author's own
records)

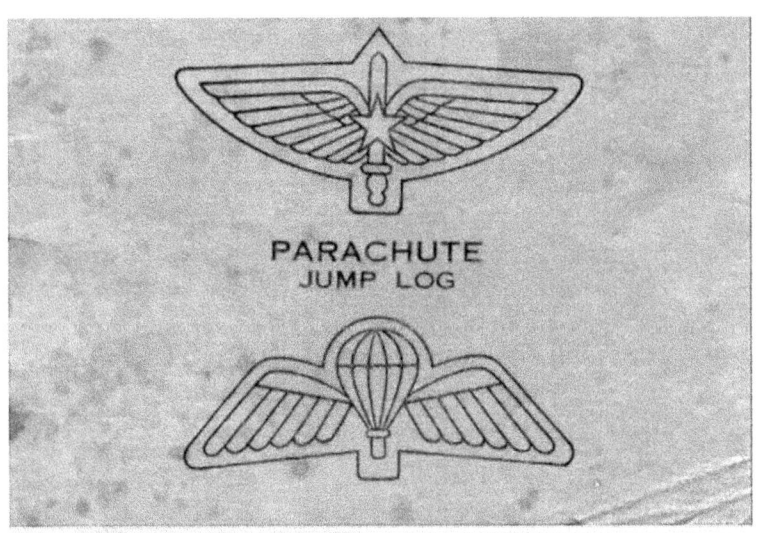

Author's Parachute Jump log book signed by late Brig. Tariq Mahmood, OC School

CHAPTER 3

Nagi Post Battle Western Front 1971

36 Frontier Force Regiment Pakistan Army—4 Para Battalion Indian Army

After graduating from the military academy, we were given about ten vacation days to visit families. My placement was to 36 Frontier Force Regiment (36 FF), based in Quetta. Frontier Force soldiers are called the 'Piffers'— *Punjab Irregular Frontier Force*, a name inherited from the colonial army history old British India. On joining the unit, I found we were six 2nd lieutenants, and I was the junior most. Had I not been punished and demoted at the military academy, I would have been the senior-most among the six officers; that is when I realized the extent of the damage, I had caused to myself by bunking classes at the military academy.

My battalion, 36 Frontier Force Regiment, was part of the 25 Brigade located in Quetta. Operationally, the brigade was the reserve brigade for 4 Corps based at Lahore. I arrived at Quetta only to find a few soldiers as rear elements while the rest of the battalion was on tactical exercises at Sibi, a place far away from Quetta city. By Oct 1971, mobilization drills for the war had already begun, and my battalion was put on special military trains and rushed to Bahawalpur, in the central part of Pakistan.

After a long train journey, we arrived at night at the
Bahawalpur railway station. All through the night, the
stores were unloaded and taken to the temporary camp
located in a large mango orchard, which belonged to the
Nawab of Bahawalpur. We stayed in the orchard tented
camp for about a month and continued preparations for
the war to come. The time was spent practicing tactical
maneuvers the battalion was expected to undertake during
the war. We practiced river crossing operations as part of
the brigade's plan for setting up a 'bridge head' to facilitate
breakout operations for the strike division. A soldier may
train for a lifetime, but only a few lucky ones experience
the real battle. My destiny tagged me to be among that
group of lucky soldiers to have experienced battle on more
than one occasion.

Sometime in November 1971, the battalion left
Bahawalpur and, after marching for four or five nights,
we arrived and camped near Bahawalnagar, a small
town about ten kilometers from the border. By mid-
November, one company of approximately 150 soldiers
was deployed forward on the border. These soldiers were
holding the defense positions right on the zero line to
prevent incursion in this sector. On December 3rd 1971,
the Commanding Officer received orders to deploy the
battalion at Jalwala Headworks on the border immediately.
The Commanding Officer presumably had an early
warning of the likely deployment because before heading
to the Brigade Headquarters, he had passed instructions
for a mobilization plan to be put in place and ammunition
to be issued. By the night of December 3rd 1971, the
battalion had occupied its defense positions on the border

with India. These positions were prepared before the war, based on plans of which I am not aware.

The closest Indian city across the border was Sri Ganganagar. The battalion held a defense position across the irrigation canal at Jalwala headworks on the Bahawalnagar border. The task was to defend against any incursion in this area. To the best of my knowledge, the battalion was not aware of the presence of the Indian Para Brigade at Sri Ganganagar. All through the war, the battalion quietly sat in the defenses, and nothing much happened except for occasional artillery gun shelling. There were a few occasions when the Indian aircraft strafed and rocketed the Jalwala canal bridge from such low heights that I could see the pilot standing on the canal bank. Most soldiers fired their G3 rifles, hoping to hit the attacking aircraft. On one occasion, I remember grabbing a G3 rifle from a nearby soldier and firing at one of the low-flying aircrafts. To me, the Indian air force attacks appeared like training runs for new fighter pilots, shooting at an undefended target within a few hundred yards across the border and rushing back to the safety of Indian air space.

My battalion was part of the 25 Infantry Brigade, which had deployed its three infantry battalions on an extensive front with inter-battalion gaps of over ten kilometers or more. Inter-battalion gaps were plugged by a chain of permanently built border posts held by a border militia force called the Rangers. The few skirmishes or battles that my battalion fought with the Indian Army were after December 16th 1971, when the war had officially ended, and both sides were observing the ceasefire agreed to by the two countries.

The first major skirmish after the ceasefire was after midnight on December 17th 1971 when Indian Border Security Forces attacked and captured a border post held by Pakistan Rangers. Rangers are a lightly armed paramilitary border security force. This border post was located on the zero line a few kilometers on the left flank of my battalion defense. Approximately thirty soldiers from the Pakistan Rangers were defending this border post. They had a field telephone line connected to my battalion 36 Frontier Force Regiment. As the attack developed through the night, the post kept reporting, but the telephone went dead sometime before daylight. We assumed that the Indians had captured the post, but no definite information was available. This post was outside the range of the field artillery guns deployed in direct support of my battalion. The morning after the post was lost, two artillery guns from the gun battery were temporarily deployed forward to bring fire to the post. This was not a good decision because, with the manual plotting practices of the time, GPO (Gun Position Officer) resources could not be split to support the two separated gun positions. GPO, with the help of his technical assistants, manually computed the firing data and provided them to the guns. One is not expected to question orders in the military, so the two guns did move but never became operational to bring fire to the border post. However, I believe such a manual target plotting process has now been automated using small portable computers. Computer-assisted automated calculations with datalink may allow separating the guns into multiple positions.

The following day, I was sent with some soldiers in two small trucks to find out what had happened to the

Ranger post. My task was to find out if the Indians had captured the Ranger post or if there were still some soldiers holding on to the position. The battalion commander expected a report for him to decide on the next course of action. In this area, the irrigation canal ran parallel to the border with no bridge for vehicles to cross. I left the two trucks on the home bank and crossed over the canal using a small footbridge. I went forward along with the soldiers taking cover along the trees. When we were a little distance away from the Ranger post, we were fired upon by the Indian BSF soldiers who were clearly in occupation of the Pakistan border post.

I do not think I made any plan or gave any kind of operational orders or verbal orders taught to us at the military academy. I simply called upon the soldiers to follow me, and we rushed forward, firing and attacking the post. The two artillery guns which were moved forward earlier were still not in communication with us, and so there was no fire support available. Being an inexperienced young officer at the time, I did not consider any of the planning considerations and battle procedures taught to us. We simply charged ahead in broad daylight like idiots, and it was like, let's go, follow me, guns blazing. I think we surprised the Indian soldiers who left the post and fled back to their BSF post that we could see at a distance of maybe 500 yards into the Indian territory.

Having recaptured the Ranger post in a style that was not in accordance with the prescribed military procedures, eventually, I did one thing right: reorganizing the post defense and putting the soldiers into proper fire positions to counter any further attack by the Indians. As I went around, I found the bodies of a few dead Ranger soldiers while others had disappeared, I guess, into neighboring villages. Even after fifty years, I can still recollect the sight of one dead Ranger soldier in a Light Machine Gun V-shaped trench. He still had his hand on his Light Machine Gun and had a big wound in his chest; he had most likely taken an automatic weapon burst from a close distance and appeared to have been firing until he died.

During the attack, we captured one Indian Light Machine Gun (LMG), called the Bren Gun. Intermittent fire exchange continued for some time. I was lucky to have escaped being shot as one of the soldiers standing next to me took a shot in the face, and I had to hold him as he went down. The injured soldier was evacuated and luckily, he survived. After reporting the situation to the Battalion HQ, I was given orders to return as somebody else was assigned to take responsibility. This area was not part of our battalion defense proper, and I am not sure who took care of this area. However, later I noticed some elements from 44 Baluch Regiment arriving to reinforce the battalion.

Nagi Post Battle Western Front 1971

The Independence war of Bangladesh had officially ended on December 16th, 1971. The second incident occurred on December 27th, 1971 well after the ceasefire, and this was a day-long battle followed by a dawn attack by the 4 Para Battalion of the Indian Army. During the war, the 36 Frontier Force battalion of the Pakistan Army had taken up the defense along the village tree line and cotton plantations almost on the zero line. As far as I know, my battalion was not aware of the presence of the Indian Para Brigade across the border in this sector. The war had officially ended on 16 December 1971, so there should not have been any incursions or large-scale battles, but such a battle happened on December 27th, 1971, around a place known as Nagi in the Bahawalnagar sector.

The episode started a few days after the ceasefire when the Brigade Commander came for a visit and instructed the battalion commander to occupy a few sand dunes which were about 200 meters in front of the main defense, but clearly inside India. On the Indian side, the village and the tree line were about 500 meters away from the border. The stretch of open sandy area up to the Indian village was no-man's land.

In this sector, the Pakistan Army had not crossed the international boundary during the war, and now, after the war, the battalion was being asked to occupy a few insignificant sand dunes across the border on the Indian territory. There was no apparent logic in doing so, and I do not understand even today the anticipated military gains of the Brigade Commander's instruction.

The intended position could, at best, be a temporary screen position, but I would not be comfortable with that also because the position was barely two hundred yards in front of the main defense and would be obstructing the direct fire of the battalions' main defense positions. Also, politically it was an incorrect decision as the ceasefire agreed upon by both governments was in place. The Brigade Commander's desire for some activity had simply exposed the battalion to undefendable vulnerabilities. The decision was to send a platoon of approximately 36 soldiers into the position, the maximum number the area could tactically accommodate. The soldiers were instructed to occupy the defense positions each day after dark, stay through the night and withdraw before daybreak, leaving only a squad of ten soldiers to hold the position during the day. The platoon would return after sunset each day to hold the place at night only.

One infantry platoon consists of about 36 soldiers organized into three squads or sections of ten men each, with a junior command officer. The Indians knew very well that these positions were not taken during the war, as we were to find out later that Indian foot patrols used to visit the area routinely. We did not want the Indians to know the Pakistan Army had occupied the sand dunes in no man's land; therefore, the position was lightly held during the day and the soldiers remained hidden in trenches.

On that fateful morning of December 27th, 1971, about midday, an Indian patrol, not knowing that the sand dunes were occupied, walked to within a few yards of the 36 Frontier Force new sand dune positions. The Pakistani soldier on LMG watch duty (*light machine gun*) instead of firing and killing on the easy target, first shouted at the Indian patrol to stop and followed with automatic fire. Since the 36 Frontier Force Regiment soldier had cautioned in a loud voice in Punjabi—*"tham, nai te goli ai hai"*—meaning "Stop or I will fire," the Indian patrol got time to take cover in the sand dunes. The Indian Army now had the difficult task of retrieving their patrol. They started pounding the 36 Frontier Force Regiment defense positions with artillery fire which allowed the Indian patrol to creep back to safety slowly. Indians continued pounding the positions almost the whole day with interruption.

We were using a portable VHF (very high frequency) radio set called PRC-10. One would rotate the dial to set the desired frequency. In my free time, I would often take a spare radio set, sneak into some corner, fiddle with the radio to search and enjoy listening to chats which was a bad practice and forbidden. Once I bumped into the frequency of the Indian artillery battery and found it interesting to listen to the radio chat of the observer.

On the morning of December 27th 1971, after the patrol incident when Indian guns were shelling the Pakistan Army defense positions, I started listening to the Indian Artillery observer's radio chat, who was sending back target map coordinates. The Indian artillery observer seemed excited and did not care to follow the radio discipline. He exposed the operational plans by using words such as "register as target 0", and "FUP" a phrase to indicate the forming-up-place for troops to assemble before the attack. The use of these words revealed their intention to attack the registered targets. The Indian artillery observer's worst mistake was when answering to someone's query, he said, "Tiger 4 is with Big Tiger in a meeting", which meant CO 4 Para was in a meeting with the Brigade Commander. Poor radio discipline and loose talk by the FOO (Forward Observation Officer, a gunner officer) on the artillery battery radio net allowed us to collect some precious information on Indian Army plans, which are enumerated below.

- Indian Army 4 Para Battalion was present opposite to the Pakistan Army's 36 Frontier Force Regiment.
- Indian Army Para Brigade was at Sri Ganganagar.

- 4 Para planned to attack the sand dune position that 36 Frontier Force Regiment had occupied after the war.
- Among the targets registered by fire were the sand dunes position and another dominating sand dune named 'Nagi Post" This was a forward platoon position of 36 Frontier Force Regiment's main defense close to the contested dune.

Commanding Officer of 36 Frontier Force Regiment immediately briefed his Brigade HQ on the Indian Army's intentions to attack and recapture the sand dune area. The 36 Frontier Force was provided additional fire support resources to counter the likely attack by the Indian Army. On its own, 36 Frontier Force had only one artillery battery of six guns in direct support, which was deployed just behind the battalion defenses. Operations people in the battalion were super excited that the Indians had revealed their intention already. Now with the newly acquired information, one regiment of eighteen guns was placed in direct support of the battalion. From the lay of the land, the Indians had very little choice on the direction for the attack. The newly allocated eighteen guns were laid targeting the Indian Army's likely assembling points. Guns were kept ready to fire, waiting for the attack to begin. Targets were registered using a procedure the gunners call map registration, meaning that the data required for firing the guns are calculated on the map. Additionally, the battalion had its own 3-inch mortars to provide fire on the likely targets. One can appreciate the effect of eighteen guns plus six mortars firing on a small target barely covering an area of a few hundred yards.

Since the Indian attack was known to be coming that night, the commanding officer decided to send the battalion Second in Command (2IC), the senior most officer after the commanding officer, to go and stay that night with the sand dune platoon. I don't know why the commanding officer took this decision, but most of us were surprised. Placing a senior officer in a lightly held, small, insignificant position was not a good decision.

I can still remember it was a night with a full moon which in the desert provides good visibility. We expected the attack to begin after the moon went away and darkness fell, which we estimated would be around 3 AM. As expected, the Indian attack started after the moon faded and the night darkness fell. The Indian attack was preceded by heavy artillery shelling of the sand dune platoon position. The moment the battle started I was allowed to open up one PRC-10 radio to listen to the Indian artillery radio. Listening to the radio chat provided the 36 Frontier Force Regiment with real-time live information on the direction and progress of the Indian Army 4 Para Battalion's ongoing attack. The moment the Indian artillery started pounding the platoon position, all the guns from the Pakistan side also opened up, shelling the small target area. Guns were bombarding the area from both sides.

The terrain was desert, and the entire area was soon covered by dust clouds choking both the weapons and the soldiers. The Indian artillery radio chat reported that 4 Para attacking Indian troops had lost direction. The situation worsened due to the growing number of causalities caused by heavy shelling from both sides. The Indian 4 Para attacking troops requested the Indian tanks to fire tracer rounds onto the target to indicate direction. I can clearly remember the call reporting that an Indian Army officer by the name Major T.J. Singh and the Artillery Observer, with the attacking troops being reported as casualties. We were to find out later that the Indians suffered very heavy losses. On the Pakistan side, causalities were few because only the forward dune position was being attacked. The dust cloud caused by the intense shelling forced sand particles inside the weapons, resulting in the weapons getting jammed and ceasing to function. Soldiers had to quickly take remedial action by disassembling, cleaning, and reassembling the weapons to work again.

During the attack, as the 4 Para company commander became a casualty, one young platoon commander was asked to take charge and continue the attack, which he did successfully. The Indian Artillery Observer had also become a casualty. However, as I could hear on their radio, his TA (Technical Assistant) had taken over the responsibility of helping with the artillery shoots. He did perform a commendable job. The Indians finally captured the position as the 36 Frontier Force Regiment soldiers, withdrew to the main defense.

A few years later, when Bangladesh became an independent country, and I was serving with the Bangladesh Army, I was sent for a weapons training class to the Infantry School at MHOW India. While attending the weapons training course in India, I met that young Lieutenant Michael Angelo from 4 Para, who had taken charge of the company that night to continue the attack successfully. We became good friends and he also took me for dinner to his old Commanding Officer's residence for dinner. Colonel Goel, the old Commanding Officer of 4 Para in December 1971 was happy to meet me and on dinner smiling told his wife, 'look here is sitting someone from the other side and we were shooting to kill each other in December 1971.'

During the early hours in the morning in December 1971, the 36 Frontier Force had lost radio contact with the sand dune platoon and had to assume that the position was lost. Interestingly the Indian Army Air OP, *an airborne artillery observer flying a light aircraft*, kept reporting on the radio on the situation. Listening to that radio chat, we learned the names of the Pakistan Army soldiers who were taken as prisoners including one injured. The radio chat even provided the Pakistan battalion with the list of items lost, like one marked map which was carried by the 2IC and one MG1A3 machine gun. After capturing the sand dunes, the Indians made no attempts to go further. The battle for the platoon position was over, but the firing and sniping continued.

In the morning around 7 AM, with the battlefront now quiet, I accompanied my Commanding Officer (CO) to the forward position at Nagi Post for the CO to see the battle situation. At one stage, unknowingly, I left my commanding officer and on my own joined the crew of a 75mm anti-tank gun, taking them forward to engage an Indian tank which had come close to Nagi Post.

The lone Indian tank that accompanied the 4 Para attack had taken up a fire position about 400 meters away from Nagi's post defenses after encountering the minefield. I had just come out of the military academy, so I had no training in firing a 75mm anti-tank gun. I asked the crew to show me how to fire. With the instructions that I received on the spot, I engaged the Indian tank firing the 75 mm anti-tank gun twice, and each time, the tank fired back at me with the main gun and machine gun. I was behaving like a kid having fun. The tank shells exploded as close as two to three meters covering me with sand dust. I would dust myself off and fire again. Eventually I was able to hit the tank and it was damaged. The tank was later recovered after dark the same night.

I must have been in the prayers of my parents because I survived two main gunshots from a tank from a distance of about 400 meters. Each time the tank saw the back-blast flash of my Recoilless Rifle gun, the tank fired back targeting me. Each time I could see the tank main gun muzzle flash followed by an explosion boom next to me. Military people with experience on battle tanks may find it difficult to believe my story that I survived a tank main gun shot from such a close distance; well, I think God had ordained a longer life for me, so I lived to tell the story today.

At about the same time, our company commander Major Saranjam, late Lieutenant General Saranjam, was injured in the trench next to me, being hit by splinters from Indian artillery firing air burst shells over our forward defense positions. One of my gun crew was also seriously injured, taking a full machine gun burst on his right arm. That was about the time when I received a message to go back to battalion HQ immediately, ending my battle exposure.

Having spent so many years in the military, in hindsight, I may say that tactically it was a bad decision to put one platoon detached from the battalion's main defense ahead, blocking the fire of the main defense itself and with no tactical gain. Years later, when I became an anti-tank training class instructor, I used to share my story with my students. In the armies I served in, including my training with the Indian Army, the teaching is to immediately move to a different fire position after firing a shot at the tank, because tank guns are not likely to miss their targets at such close range. Here I stood to defy that theory to have survived two main tank shots from a distance of barely 400M.

The war ended on December 16th, 1971 with East Pakistan breaking away to become an independent country known to the world as Bangladesh. Overnight we transitioned from being citizens to being non-citizens in Pakistan. In February of 1972, Bengali officers like me were taken out from the outfit of the Pakistan Army and moved into repatriation camps as we were no longer citizens of Pakistan. While I was in the repatriation camp, I learned through the local Pakistan radio news that I had been awarded the highest gallantry award given to a living soldier. The other higher gallantry award given only posthumously, and therefore my award, "Sitar-e-Jurat— *Star of Valour*" is considered the highest for a living person. I never received any official notification of the award from the Pakistan Army. As a young officer, I risked my life for an army that locked me up soon after the war in so-called repatriation camps for two years. The Pakistan Army even rejected my petition to allow me to stay with my parents who were at the air force officer" repatriation camp at Warsak camp, barely two-hour drive away from my Camp.

TOP: 36 Frontier Force
Regiment, Pakistan Army at
Bhawalnagar, 1971. Author at
far right (Source: Author's own
records)

MIDDLE: Farewell Lunch. 36
Frontier Force, Pakistan Army.
Bhawalnagar, 1971 (Source:
Author's own records)

RIGHT: Sitare Jurrat (Star of
Valor) award by Pakistan Army
for Naggi Battle Bhawalnagar,
1971 (Source: Author's own
records)

CHAPTER 4

Prison Camps & Repatriation to Bangladesh

In December of 1971, Bangladesh became an independent country; and in February of 1972, Bengali officers who were serving in the Pakistan Army and had opted to return to Bangladesh were taken out from their serving locations and taken to different repatriation camps. The war ended with thousands of Pakistan Army soldiers who had surrendered in Bangladesh being detained in India as POWs, prisoners of war. Therefore, the repatriation of the Bengali military personnel became uncertain. Bengali non-military or civilian families continued to stay under their own arrangements in different cities, mostly in Karachi. However, at the time, Bengali military officers serving in Pakistan and their families were taken and interned in repatriation camps at various locations across Pakistan. I use the term interned because Bengali military officers and their families were restricted within the camp perimeters and not allowed to go outside, not even to buy the essentials like groceries. Movement restrictions grew tighter and tighter over time. Repatriation camps were created at old military facilities, and perimeter security was set up to prevent the Bengali military personnel and families from going outside. I left my unit, 36 Frontier Force, which was deployed at the time on the Bawalnagar border and reached the army garrison at Okara Cantonment. Over the next few days, dozens of Bengali officers arrived at the Okara collection center. Subsequently, the Bengali officers were put on a special train and taken to different repatriation camps. I was with a group of five hundred army and air force officers who did not have families with them. We were put on a special train that took us to Shagai Fort, located in Khyber-pass, close to the Afghan border.

My parents and siblings ended up at Warsak Repatriation Camp with other air force officers and families. Warsak Repatriation Camp was located near Warsak Dam near Peshawar, about a two-hour drive from my location, Shagai Fort repatriation camp. Shagai Fort has a fascinating history. It was built by the Colonial British Army in 1927 to station military forces to guard against Afghan invasions into India. One may like to read more about Shagai Fort, located in "Kyber Pass" Pakistan, as it has a fascinating history connected to Indo-Afghan Wars. Shagai Fort was built more than a hundred years ago in order to house the local militia force to fight the Afghan invaders, so the Shagai Fort accommodation had just basic amenities. Like the old-time fortresses, on the outer side the rooms had no windows except for a few small openings in every room with heavy metal sliding doors. These openings were for militia soldiers to fire back in defense if the fort came under attack. The room entrance doors opened facing the inner courtyard. There were possibly a few rooms with proper amenities, which were being used by Pakistan Army officers and staff deputed for our security.

The fort had only one large entry gate, which was colossal, like one would see in any fortress. This gate always remained closed, and we were not allowed outside.

Local armed tribal militia guarded the fort; they were assembled in small groups in tents outside the fort. This may or may not have been true, but we did hear from local employees that the tribal militia outside guarding the fort were told that Indian military prisoners were being held inside, so they should shoot anyone trying to escape. They did indeed shoot one officer dead when he jumped outside from the rooftop, possibly in a suicide attempt. Officially these were called repatriation camps to house Bengali officers who had earlier served in the Pakistan Army and had even fought the war for them. Restrictions and living conditions in these repatriation camps were atrocious and more like POW camps. Just like in a prisoner of war camp, every morning, we were required to stand in groups for the headcount by the camp in charge to make sure nobody had escaped the night before.

In Shagai Fort, a room was allocated to a group of ten to fifteen people, numbers being dependent on room size. These rooms were made for the tribal militia soldiers without any kind of facilities like fans or electrical outlets. The rooms were like a hollow spaces into which a dozen beds had been thrown. Initially, a few dozen Bengali soldiers were brought in and assigned to look after the officers, the 'batman' service of the old army. These Bengali soldiers had earlier been withdrawn from their old units and were waiting for repatriation like us. These Bengali soldiers gradually became reluctant to work for their Bengali officers, so they had to be taken out and moved to repatriation camps exclusive for soldiers.

With life restricted to rooms and a small courtyard in the middle of the fort, officers adapted to the prison-like lifestyle by keeping busy and passing the time. People formed groups with a common interest and spent their time playing cards, listening to music and occasionally playing basketball or squash. The fort had one basketball court and one squash court. We were provided food by a local contractor. At breakfast, lunch, and dinner time, we would all gather and stand waiting outside the dining hall. After laying out the food on the dining tables, contractor employees would open the entry door. Food quality was sufficient for survival needs, but nothing to look forward to. Over time, almost in every room, one officer became an amateur cook for the occasional room cooking. We were provided a small monthly pocket money, and the amount was about Rupees 35 a month *(1972 1 USD=11.01 PKR Monthly Pocket Money of 35 PKR = USD 3.18)*. With this we could buy essentials from the shop run by a local tribal person inside the fort exclusively for the interned officers. This shopkeeper made good business selling essentials like biscuits, other eatables, warm jackets and sought-after electronic items at highly inflated prices. He collected required item lists from officers and managed to smuggle inside everything requested. He must have been paying good baksheesh (tips) money to the camp security people. This person made tons of money, especially selling electronic devices like radios and tape recorders to interned officers who spent time listening to popular songs and music. Some senior officers had cash from personal savings and were happy to loan some money to junior officers. Loans were given on the premise that these would be repaid later on return to Bangladesh; and frankly, it's unclear if these loans were ever paid back. My parents were in Warsak camp, which housed the air force families. From my Shagai Fort camp, Warsak camp was at a drive distance of about two hours, but I was not allowed to go and meet my parents even on the annual Eid day. My petition seeking permission to stay with them was rejected twice by GHQ (*General Headquarters—the Pakistan Army Headquarters*).

After two long years of internment in the repatriation camps, the repatriation process started under arrangements of the **ICRC**—*International Committee for Red Cross*. In November 1973, a month before the start of repatriation, I was finally permitted to join my parents in the Warsak repatriation camp. We were among the first families scheduled for repatriation from Warsak Camp. We were put on a special train at Peshawar and taken to Karachi. In Karachi, we stayed at the air force base at Drigh Road for a few days. Finally, I, along with my parents and siblings, were repatriated on 23 Dec 1973 on ICRC chartered flight of Ariana Airlines—Afghan Airlines. We landed at Tejgaon airport in Bangladesh. This was the only airport for international flights at the time. Going down the aircraft ladder, and stepping onto the soil of our motherland, the newly independent country for which so much blood had been shed over the past year was truly a moment of great jubilation for all of us. Finally, we had returned to the homeland we left years before in pursuit of a job.

Our repatriation also meant the end of two years of prison life, and it meant our freedom. Pakistan used the Bengali interned personnel in various camps as a bargaining chip to get their prisoners released from India, thus the two years of detention for us. Pakistan had agreed to let us go only after India agreed to repatriate the Pakistan Army soldiers held as prisoners of war in India.

At the airport, we completed the documentation and went home. My Father had left his village home in 1944 and served all those years away from home in Pakistan. We were all born and brought up in Pakistan, so we were like refugees returning home with few belongings and no money. To our relief, we had some relations who came to receive us and took us to their home. Like a refugee family, we had the difficult task of rebuilding from scratch. After about a week, my father joined the Bangladesh Air Force and started working with Air Force Records Office at Kurmitola, Dhaka. I went to the special setup army reception center, where I received my placement orders. I received one month's salary in advance, which was about Taka five hundred. This small amount was a huge relief for people like me making a new start in Bangladesh.

There would be many readers who have never lived in Bangladesh and may not be aware of its turbulent history and how it split apart from Pakistan to become an independent nation. So, for the benefit of readers who are not from Bangladesh, it would be helpful to write a few lines of history explaining very briefly how and why Bangladesh emerged as a new independent state. These introductory history notes may help the reader to relate in the proper perspective the issues mentioned in different sections of the book. This may also help to understand why the Bengali military officers in Pakistan were interned and locked in the repatriation camps for two years.

The British left India in 1947, partitioning the country into two halves based on religion. Hindu-majority areas became India and the Muslim majority areas became the new country called Pakistan. 'West Pakistan' and 'East Pakistan' were carved out of India with Muslim majority areas to form Pakistan. Thus Pakistan had two separate wings separated by approximately a thousand miles of Indian territory in between. People in West Pakistan and East Pakistan spoke different languages, had completely different culture, had different food habits and even dressed differently. Around eighty percent of the population in East Pakistan were Muslims, the rest of the population were Hindus, Christians, and a small number of Buddhists. Religion was seen as the only common element binding the two wings, a perception which would be proved wrong years later. While religion has been used throughout history in an effort to unite people under a common flag, history tells us that ethnicity, language, and culture are more dominant factors, especially when it comes to nationhood or national identity.

The freedom struggle, and the struggle for Bengali identity has a long history. In the new nation of Pakistan, Bengalis rejected the Pakistan government's demand to accept 'Urdu' as the official language in this part of the country. Everybody in then East Pakistan spoke Bengali except for a small group of people who had arrived from India after the 1947 partition and had settled here. The language thus became a driving agenda in the political struggle for the independence of Bangladesh. Along with the language issue, years later economic deprivation also caught up with the people's demands. After the national elections in December 1970, the governing elites from West Pakistan were shocked to see that Awami League, the political party from East Pakistan had swept the polls. They refused to hand power over to the Awami League, which had won majority of the national parliamentary seats. Political negotiations failed, and the Pakistan Army launched a major crackdown on the night of 25 March 1971 in East Pakistan to suppress the political uprising. The next day on 26 March 1971, Sheikh Mujibur Rahman, the Bengali political leader who had won the elections, declared the independence of Bangladesh, and called for the Bengali nation to rise and fight for independence. Bengali soldiers stationed in East Pakistan rose to the call and formed the resistance army for the independence struggle, thus the Bangladesh Army was born. The same day Sheikh Mujib was arrested by Pakistan Army and flown away to Pakistan to be imprisoned for the next two years. To the people of Bangladesh, Sheikh Mujib popularly known as 'Bangabandhu' meaning Friend of Bengal, is recognized as the Father of the Nation. During that time, the Pakistan Army launched a hunt for Bengali political leaders and their supporters, which led to the mass killings of civilians. As a result, many Bengali political leaders were forced to flee to India to seek refuge. In the absence of Sheikh Mujib, other political leaders organized the independence struggle. Bengali army officers and soldiers stationed in East Pakistan revolted and formed ranks to create the core element of the new Bangladesh Army. Around this core element the numbers grew and waged the liberation war successfully. People from all walks of life, especially thousands of students, joined the liberation force called the 'Mukti Bahini—*Liberation Army*. After about nine months of war, the Pakistan Army surrendered in Dhaka on December 16th 1971 to the allied command of Bangladesh Liberation Forces and the Indian Army. The eastern part thus achieved its freedom as an independent country to be called 'The Peoples Republic of Bangladesh.'

CHAPTER 5

Bangladesh Army—Turmoil in Formative Years

In the newly independent country, the biggest challenge for the new army in its formative years was to forge discipline as the military transitioned from a wartime force to a regular army. This was a challenging task, and it took years to reign the soldiers into disciplined military life. In the military, typically, a soldier's life is regulated; we live simply on orders and instructions for the day till we hit bed at night. I believe most countries that have experienced a civil war have had such transitional problems in the military. In its formative years, the new Bangladesh Army had to deal with multiple frictions and divisions, particularly within the officer's corps. A major issue was the divide between officers who had fought the Liberation War, putting their lives in harm's way and the repatriated officers like me who were stranded in West Pakistan and did not participate in the Liberation War. The other more severe issue that plagued the Bangladesh Army in its early years was an effort of subversion by leftist political elements to politicize the soldiers.

In the 1970s, socialist ideologies and the concept of a classless society gained significant ground in countries like Bangladesh. During this time, certain political parties in Bangladesh advocated for the elimination of a distinct officers' class with special privileges. These parties propagated the belief among soldiers that there was no necessity for a privileged officer class, as soldiers were equally capable of assuming leadership and managerial roles. Consequently, a wave of class struggle ensued throughout the 1970s, leading to a series of mutinies within the army. These mutinies resulted in officers facing humiliation and, in some instances, being shot and killed by the soldiers.

Japan Airline Hijacking and 1977 Mutiny

In Oct 1977, a Japanese Airline passenger plane was hijacked by the Japanese Red Brigade members. The aircraft landed in Dhaka and was parked at the old Tejgaon International airport. Many senior air force officers were present at the Tejgaon airport, with some staying overnight at the air traffic control tower negotiating with the hijackers. With the nation's attention focused on the airport and the hijacked plane, a group of air force soldiers, primarily from Kurmitola air base, took up weapons and started a mutiny.

Their primary objective was to get rid of the officers. These soldiers were made to believe that they could take up the positions and responsibilities held by the officers.

The armory was forced open at night, and the soldiers stole weapons. Many army soldiers also joined the mutineers from nearby locations in Dhaka. By next morning, the mutiny had spread all over the Dhaka garrison. Dhaka city was in total chaos throughout the night, with killings and looting happening up to the following day.

The presence of senior air force officers at Tejgaon International airport presented an easy target for mutinous soldiers. The next morning around 7 am, a few soldiers arrived in a truck at the gate of Tejgaon air force officers' mess (residential unit) opposite to the airport. My father was serving in the Air Force at the time, and the family, including my siblings, were staying at the family accommodation inside the officer's mess premises . In the morning of 02 Oct 1977, the soldiers started firing indiscriminately towards the family accommodation. My youngest brother Enam, was sitting next to the window in our house when a bullet came through the window and hit him on the head, killing him instantaneously. It's a trauma that still pains and haunts my family and me.

My brother was shot because he belonged to parents representing the officer's corps. The night before, six senior air force officers were also shot dead at the Tejgaon airport control tower building by the mutinous soldiers for no reason other than that they wore the officer's rank. Political subversion had driven these soldiers to simple madness and seven needless deaths resulted. It was a late afternoon with a gentle drizzle, as I remember my young brother and six air force officers were buried at Shaheen School graveyard in Dhaka.

The following day, the army regained control, and the mutinous soldiers were arrested, although some managed to disappear. The arrested soldiers faced trial by military court martial and were subsequently handed either death sentences or lengthy prison terms.

Over time, several such incidents occurred and many mutinous soldiers were sentenced to death and hanged, while others were given prison sentences. As we reflect on these punishments half a century later, there is ongoing debate about their harshness. The memories of mutinous soldiers engaging in acts of violence, killing, and looting have faded from our collective memories. We now are living in stable and peaceful conditions and the smell of blood and death is absent.

Given these circumstances, it can be challenging for people to fully comprehend the profound fear experienced by those who faced the imminent threat of being shot or subjected to humiliation at the hands of unruly armed soldiers. The drastic measures taken at the time may seem excessive in retrospect, but it is important to recognize that they were implemented to prevent further mutinies and the associated instability for the new nation as a whole.

As time has passed, it has become evident that these severe sentences ultimately put an end to the cycle of mutinies and the accompanying violence. It is possible that the mutinous soldiers themselves came to realize that they were being manipulated by external political forces. They began to understand that their actions would lead to their own suffering, as the mutinies were short-lived, lasting only a few days, and did not result in the fulfillment of their demands. While the severity of the punishments may be a topic of debate, their effectiveness in quelling the mutinies cannot be denied. From the early 1980's discipline slowly returned, and the army became more stable. I was stationed out of Dhaka during most this tumultuous period, so fortunately, I was always out of harm's way.

CHAPTER 6

Service with Bangladesh Army

After repatriation, I joined 5 East Bengal, an infantry battalion, at Comilla in Jan 1974. During the 1971 war, 5 East Bengal was deployed for operations on the border in West Pakistan when a large number of its soldiers defected and walked across the border into India, hoping to return to Bangladesh quickly. After this incident of soldiers' defection, 5 East Bengal was disbanded as a military unit; its remaining soldiers were disarmed and interned in camps. These interned soldiers were allowed to return to Bangladesh after about two years. 5 East Bengal was re-raised in Bangladesh in Comilla in Jan 1974.

The raising ceremony was a historical event as the regimental flag was raised at the Quarter Guard by Bangabandhu, Sheikh Mujibur Rahman,' Bangladesh's Prime Minister. On that afternoon, a lunch was arranged in honor of Bangabandhu, Sheikh Mujibur Rahman, at the hilltop officers' club, where all the garrison officers had the opportunity to meet Bangabandhu in person. This historical event may have missed being recorded in the unit's history as the current generation of officers in 5 East Bengal need to be made aware of this important event.

This flag-raising event was a unique honor for 5 East Bengal, which at the time had Major Abdus Salam as the Commanding Officer who years later retired as Major General. In 1975, Comilla garrison had only one brigade with Colonel Huda as the Brigade Commander, who was killed in Oct 1977 in Dhaka after a coup which also triggered soldiers mutiny (*sepoy mutiny*). Such mutiny and counter-mutiny in the military were frequent at the time. The re-raised new unit had barely settled on rebuilding itself when it was assigned the duty of anti-smuggling operations in Sylhet Srimangal tea garden area. The unit was deployed in small groups over an extensive area along the eastern border of Sylhet. Starting in the north at Baraleka and going south along the border to Kulaura, Shamshernagar, Srimangal tea belt and on to Teliapara in the south.

The Battalion HQ was located at Srimangal, housed at a government facility little south of the city. At the time, Srimangal was a very small township compared to what it looked like in 2022. In 1975 the military units had very few transports of their own, especially small cross-country capable vehicles having all-wheel drive. Since this deployment was 'In Aid of Civil Administration', the civil administration provided the required civil transport and fuel from local sources. During our year long stay, we made new friends among the tea planters and garden managers. Tea gardens had good club facilities where the planters and their family members would come for tennis, billiards, or social get-togethers on the weekend. We were often invited to such club gatherings, which provided good recreation. In 1974, road communications from Srimangal towards Shamshernagar and further north were connected by only dirt roads. In the monsoons, driving on these dirt roads was difficult without having all-wheel drive, especially on the road strip running through Bhanugach forest. All that is now history as most roads in the Srimangal tea garden belt are now asphalt, all-weather roads. After about one year on the Sylhet borders, 5 East Bengal returned to Comilla garrison by early 1975 and got down with routine peacetime activities.

While serving with 5 East Bengal at Comilla in September 1975, I was sent to India for the Battalion Support Weapons course (BSW-9). I arrived in Kolkata, India early Sep 1975 along with another officer heading for another training course in that same city where I was going. In Kolkata, I stayed at the Bengal Area Officer's mess at the historic Fort Williams, also the Headquarters of Indian Army Eastern Command. Fort Williams was built in 1781 by the 'East India Company' on River Hoogly. The fort complex covers an area of approximately 70 acres. No wonder British Colonial authorities had constructed the fort close to the river, as their power base was their naval power. British naval power was strong at the time and instrumental in their successful colonization of many countries.

I stayed at Fort Williams for three days as I waited for the military discounted tickets which were provided to me by courtesy of the Indian Army. My train journey from Kolkata to Mhow, a small city near Indore, Madhya Pradesh, took more than 24 hours. It was a long journey, and was not comfortable. However, I was young at the time, maybe mid-twenties, so I had no complaints. The Battalion Support Weapons Course I attended comprised of three separate training modules combined into one training class of five months. The three training modules were: 1) an anti-tank training module based on 106 mm Anti-Tank Recoilless Rifle; 2) Battalion Mortar training module based on the French 81mm mortar; 3) 7.62mm Machine Gun training module. My knowledge of infantry weapons had a good foundation as a result of my weapons training in India. To me, the big difference between Pakistan military academy weapons training and weapons training in India was that the focus was on teaching, not ragging or physical punishments. I did fairly well on my course and was recommended to be a weapon instructor, a job I would perform later in Bangladesh Army.

During my stay in India from September 1975—January 1976 there were multiple military coups and soldier mutinies in Bangladesh. These coups were a result of power struggles between the military leadership and compounded by political subversion of the soldiers who had the agenda to get rid of officers. During the November 7th coup, several senior and brilliant army officers were killed. The political situation in Dhaka city looked very grim, and many officers went into hiding and avoided going to their military establishments where soldiers were out to kill officers.

I was attending a training program in India during that period of the coups and countercoups in late 1977 and so remained out of harm's way. While in the training school in India, I remember some Indian Army officers reacting to the coups in Dhaka in a very unfriendly way and commenting, 'Salim, we are coming to Dhaka again and will meet you at Hotel Intercontinental. I also heard from other Indian friends that their units were ordered to stand by for likely operations in Bangladesh. Other overseas trainee officers did tell me that they have heard from their Indian army friends that Indian Army was standing by to venture into Bangladesh because they thought external forces had removed the government in Bangladesh that the Indian government supported.

Fortunately, good sense prevailed, and the Indian Army never undertook such ventures. I returned from India in January 1976, and by that time, the political situation had settled. I rejoined my unit, 5 East Bengal, in Comilla and worked as an adjutant till mid-1976, when I had to move to Combined Arms School Jessore as an instructor of weapons.

Combined Arms School Jessore

In July 1976, I started working at Combined Arms School Jessore as an instructor of weapons. In the initial years, Bangladesh Army did not have separate schools for different arms and services that we see now. After independence, Bangladesh Army had set up a single training institution called Combined Arms School. This single institution had separate wings like the Infantry Wing, Artillery Wing, and Engineers Wing, and they were all supported by a common pool of resources. However, in 1979 Combined Arms School was closed to make way for independent schools for each arm. Infantry Wing moved to Sylhet in early 1979 and became the Infantry School. Other arms schools also moved to new locations where they are settled now. An instructional job in the 70's was a challenging and difficult task because the Army had very few published teaching materials or GS Pamphlets (*military-published teaching books*). Instructors had to contact different sources on their own to collect the needed materials. The main source was other officers who had returned from attending courses abroad and had personal notes or publications of that particular Army.

As a weapons instructor, I relied on my collection of publications and notes from my course in India. Some of the student officers were kind to share the publications they had in their collections.

I was an instructor for heavy weapons like Medium Machine Guns, Battalion Mortars and Anti-tank Recoilless Rifles. In the autumn of 1976, the first OWJTC-1 (officers' weapons and junior tactics course) was scheduled to be run at the Combined Arms School Jessore, but the school did not have any instructor for small arms. So, I was given the additional responsibility of organizing and running the OW portion of OWJTC-1. The officers' weapons course, was of four weeks at the time. Physical and mental fitness is an integral part of weapons training. Student officers had to undergo a 9-mile run in battle dress for the physical fitness test. Students were graded based on the time they took to finish the run. So, the trainee officers, especially the young ones, would give extra effort to be in the lead group to score the full ten marks. That afternoon, we dropped off the students on Benapole road, from where they had to run along the specified route, finishing at the school inside Jessore cantonment. I followed the students on an old-time M-38 jeep, going up and down and watching the officers run. There was very little traffic, so running on the roads was not a problem. A little short of the finishing point, the lead officer dropped unconscious. He was rushed to the hospital, but sadly the officer died due to over-exhaustion. The lungs could not keep up with the surge in body demand for oxygen, and the person collapsed. Among the officers who had just completed the first weapons course, we selected three officers to become instructors for the subsequent courses. I returned to heavy weapons teaching and concentrated entirely on organizing the first heavy weapons course, 'Company Heavy Weapons Course -1 or CSW-1 as it was named.

I was the kind of person working on experiments and innovations, sometimes taking risks not called for by duty. In the 70's, the Bangladesh Army introduced 82mm mortars from China for the infantry battalions. The old soldiers knew how to handle the 3-inch mortars, but could not plot the targeting data required to fire the Chinese 82mm mortars. The mortar itself was relatively easy to fire but getting the mortar bombs to reach the desired target required firing data to be set on the mortar sights. The process of working out the targeting data is called plotting, a process which gives the mortar crew's elevation angle, deflection angle and propulsion charge to make the mortar bomb land on the target. The procedure used with the old 3-inch mortars could not be applied to the new 82mm mortars simply because the plotter, the mortar sight and the firing tables differed. I was told that in China, Bangladesh officers were only taught to take direct fire with 82mm mortars and plotting required for the indirect fire was not taught. Using my knowledge of the Indian 81mm mortars, I modified the mortar fire plotting procedure and made a modified plotting board for the 82mm mortar. To test the accuracy of the new plotter and the new plotting procedure, we went for a live-fire test at the firing range in Hathazari, Chittagong. I was provided a platoon of six mortars at the Hathazari firing range for the test. Over two days, I tested the modified plotting procedure and the new plotter using live ammunition. During the trial firing, I further fine-tuned the plotting and laying process to get an acceptable level of accuracy for the indirect fire. The new procedure was approved by the Army and adopted as a standard firing procedure for the infantry battalions. This was one contribution from my side to the Army, and I believe this procedure is still in use in the army. However, I have to say over the years, computer-assisted technology has progressed a great deal, making life easier in all walks of life. Firing accuracy depends on the accuracy of the firing data like the elevation angle, azimuth angle, and the propulsion charge to be applied. This job can easily be done by a small, rugged handheld computer, I hope the army may well have adopted computer-assisted plotting by now.

While teaching the use of the anti-tank gun in the mid 1970's, the school did not have dummy cartridges to practice loading and unloading drills by the weapons crews. At the time, we were using 75mm anti-tank guns. It was on a Hathazari firing range; after finishing the student firing for the day, I sent the students and my instructors away at a safe distance as I took an HE (*high explosive*) cartridge and then slowly began to disarm it. I first removed the fuse by cutting out the punch marks and unscrewing the fuse. Removing the fuse was the most dangerous part as the fuse is sensitive and can detonate. I took a small breather, after which I went a step further, removing the percussion cap and the tube holding the explosive train in the cartridge. I was doing all this on my own; had I triggered a detonation, I would have been blown to pieces and certainly would not be living to write the story today. In simple terms, I was being stupid, taking unasked risks to my life. Fortunately, all went well, and I managed to defuse and disarm two 75mm shells which were turned into drill cartridges. I know I would have been severely punished if the shell had exploded' It could have also killed or injured other people nearby.

For live firing practice of weapons like Medium Machine Guns, Battalion Mortars and Anti-tank Recoilless Rifles, we travelled by special train from Jessore to Hathazari firing range in Chittagong, a journey of two to three days. After every six hours of travel, the train would stop at some railway station to allow the old-style wood fire to cook outdoors next to the railway tracks. On the first Company Support Weapons Course on the way to Chittagong, we stopped at Mymensingh Railway Station to cook meals. As the kitchen staff got busy with their old-style wood fire cooking, the student officers decided to visit the home of one of their colleagues in Mymensingh city. We hired some twenty-plus rickshaws and visited the officer's parental home. We enjoyed great hospitality at the officer's parental residence. In 1970's and 1980's there was no railway bridge to cross the huge rivers, so the entire train had to be loaded onto the river ferry for the crossing. The big rivers like Padma and Jumana are a few kilometers wide, so crossing them required large ferries, which are like small ships. Loading and unloading the train onto the ferries is a fascinating thing to watch. The train halts a little distance away from the ferry site, and the bogeys are decoupled into packets of three carriages, that being the length of the railway track on the ferry barge. A shuttle engine pulls each of the three bogey packets away from the main train and slowly pushes them onto the ferry rail track.

The railway tracks and the ferry tracks are locked in alignment to prevent sideward movement. After loading, the shuttle train moves away to bring another packet of three train bogeys and the ferry is winched sideways to align the next set of vacant rail tracks with the rail tracks for loading. The shuttle engine goes back and forth, repeating the process until the whole train is broken into three bogey packets and loaded onto the ferry. It took about three to four hours to complete the loading. For the ferry to cross the river, it took another two hours. After crossing the river, the entire loading process is run in reverse. The bogey packets are unloaded and rejoined to form the original train. Unloading and rejoining takes another three to four hours. A new engine on the other side shuttles into position to pull the train onto its journey to Chittagong or whatever is the next destination. You are looking at eight to ten hours of river crossing time. So how do you spend time for such a long wait, play cards, gossip around, or walk around the small riverside hotels to enjoy the fresh river fish meal. If this was during summer, the heat and humidity would be unbearable. Goods trains followed the same time-consuming process, but for passenger trains, people would get off the train at the riverside arrival station and rush to find a seat on the waiting passenger ferry on the river. Seats were open for the first person to arrive and occupy them; the seat reservations system did not work. After the passenger train crosses the river, which could be two hours; passengers would rush to get onto the waiting train on the other side to find a seat. Some people would pay extra to have someone go ahead to occupy and hold their seats until they crossed over. Fortunately, things have changed for good; we now have the road and railway bridges over both these super large rivers, the Padma and Jamuna rivers.

After serving one year as an instructor of weapons, I was promoted to the rank of Major and was adjusted within the school to work as an instructor for junior tactic courses. During my stay at Combined Arms School Jessore, I got married on 03 March 1978. In terms of accommodation, at best, married officers could expect at the time was a tin hut family accommodation with an abundance of rats and frogs inside the house. My tin hut house was surrounded by tall sun grass, home to jackals. After sunset, the jackals would come and sit in front of the front door like guard dogs, and we were scared to open the door. I later discovered that the security lights attracted giant beetle-like insects, which after buzzing around the light bulb, would drop on the floor. This is what was inviting the jackals. They would be waiting to eat these large meaty insects, free meals with little effort.

Staff College Mirpur

In February 1979, I moved from Combined Arms School Jessore to my new workplace at the newly established Defense Services Command and Staff College at Mirpur Dhaka. Bangladesh Staff College was established with technical assistance from the UK Ministry of Defense. A team of British military officers who were designated as instructors assembled at the UK Staff College Camberley under the designated first Chief Instructor, Brig Gibson. They picked the required teaching materials and arrived in Dhaka. This team was referred to as the BMAT—*British Military Advisory Team*. Staff Course in most countries is generally ten to twelve months long. Bangladesh Army had a backlog of officers who could not do staff college, so the plan was to run four condensed short courses of six months duration to clear the backlog. These first four courses were kind of mini UK Camberley Staff Course run at Mirpur.

I worked with the coordination staff, printing and distributing teaching material among the students. Gradually the Bangladesh military officers started taking over the teaching responsibilities, and the British Army officers departed, leaving only the Chief Instructor and Senior Instructors of the Naval wing & Air Wing. I did my staff course in 1983, and after a short tenure in unit command, I returned to serve as an instructor in 1986-87. As more Bangladesh military officers joined the Staff College, the course content gradually transitioned to Bangladesh military content. During these transition years, the teaching staff had to re-write the teaching materials, presentation scripts and exercise papers. That was about the time when we introduced the 'doctored maps' for indoor map exercises. I served on staff at the staff college for about three years.

In 1979, Staff College did not have family accommodations. So, the staff and students lived in rented houses in the Mirpur Pallabi area. Living in a rented house in Pallabi was under marginal living conditions. We suffered from a lack of regular water and electric supply. However, a year later, the first residential building of the staff college was ready, and I was lucky to be among the few who were able to move into the comfort of military accommodations. My eldest son, Ehtesham, will probably remember the stories of ghost visits in that new building. In this new building, around midnight, we would hear strange sounds like somebody running on the rooftop. The story that went around was that this building was built on the site of a Hindu crematory ground, where dead bodies were cremated before, so we were receiving visits from the souls of the dead. To go to the office from our housing area, we used to walk across the small valley, which is now a big lake with a footbridge.

I left Staff College in 1982 and joined 38 East Bengal in Bandarban, Chittagong Hill Tracts. Bandarban is an operational area where the units were engaged in counter-insurgency operations. After a very long time, I was returning to a field unit, and my physical fitness was not good enough for the rigors of hill tract counter-insurgency operations. People who knew me well were waiting to put me to the test. Soon after arriving in the new location, I was sent for an orientation visit to the deep hill jungle camps. Chittagong Hill Tracts is tropical hill jungle terrain which gets heavy rains during the monsoons. For camp visits, you walk from your campsite to meet or link midway with the patrol from the destination camp.

I struggled going up and down the hills because my fitness was not up to the mark. I struggled all through my journey. I reached the first camp, Bagmara Camp with rest breaks in between my walks. I realized that walking uphill is indeed very tough both on muscles and breath. I had to drink bottles and bottles of water with salt to prevent dehydration. After lunch and a bit of rest at the first camp, we continued towards the next camp, called 'Liragoan'. This camp was another three to four hours of walking. After about two hours of trudging along a ridge line, my patrol mates pointed to the destination camp at a distance in the valley. With a view of the camp in the distance, I was assured of a much-wanted meal and rest, but I got scared when my patrol mates showed me the way down. We were required to climb down a wooden tree log ladder literally hanging onto the cliff side with a drop of about fifteen to twenty feet. These wooden log ladders are fixed by tribal people on frequently used tracks. Small steps were cut on a long tree log and the log was somehow tied to the hill side cliff. For tribal people, including women and children, it's a normal to climb up and down such cliff ladders often with a load on their heads. But it's scary for people like me who have never used such a ladder before. Anyway, with some help I did manage to get down and reach my camp where I stayed for the night. Next day, we followed a similar walk for a few hours, reaching my final destination at Benchari Camp.

I remember the name of Saif who was in charge of the camp because he had an interesting lifestyle being cut away from human habitation. He was a chain smoker and had his way to keep his cigarettes dry and dampproof. A kerosene lantern issued by the military, called hurricane, was always kept lit in his room and his cigarette pack was placed on top of the metal cover for the heat to keep the tobacco dry. During my stay at the camp, I learnt a lot about tribal life. Tribal people lived on whatever the forest provided, for instance, they would eat the python meat and sell the skin. They would also sell whatever different types of wildlife they could trap. Now this officer at Benchari camp had bought a bear cub and was nurturing the cub as a pet in the camp. When I arrived, the bear was fairly grown, and I noticed that it had adapted to eating soldiers' food from kitchen like rice and roti including tea. Years later, Saif gave the grown-up bear away to a zoo in Dhaka.

After I returned to Battalion HQ at Bandarban, I was sent on a few more counter insurgency operations. I took a patrol into the deep jungles east of Lohagara Police Station.

We camped for a few days and collected information from locals on insurgent movements in the area. However, we were not so successful in making contacts. My last long-range patrol was at a place called Eidgah on the way to Cox Bazar. We were to raid a suspected insurgent location in Hatidath Chara. We walked overnight into the deep jungle, but missed the insurgent group who were collecting taxes in the area. Here we found in the bushes lots of stacked elephant meat skewed and barbequed on bamboo sticks for preservation. I was told burning the top layer of meat helped its preservation for long periods. Insurgents would boil these almost dry meat lumps and eat them as a protein source whenever they visited the area. We also found two elephant tusks, which were later deposited with the unit, though some other stories went around that I had taken them into my personal custody. For security reasons, we would walk only during the nights and rest during the day. It was during this patrol at Hatidath Chara, about midnight as we were walking in total darkness along a shallow water stream, that we suddenly heard an extremely loud shrill scream of an elephant right in front of us. To escape the onrush of the elephant, I jumped towards the side bushes to take cover and so did the other soldiers. We realized that this was a lone elephant drinking water at midnight, and was surprised by our human presence. We were the uninvited guests in the jungle in the middle of the night. Our soldiers were very well disciplined on fire control, so nobody fired at the elephant which ran uphill away from us. In its panic the elephant could have charged into us causing many casualties. Fortunately, we had only a small case of trampling injury to a soldier as the elephant stepped on the soldier's hand.

Forty years on, I am told that most of those remote locations are now linked with roads, thus making them easily reachable now. As part of counter insurgency operations, the Bangladesh Army was also engaged in the development of infrastructural works to remove the remoteness of the tribal villages and bazars. Easy and quick access to local places, also helped us to successfully fight the insurgency. As part of the pacification program, the army was actively involved in setting up of schools and building new roads. Tribal villages in proximity to army camps also enjoyed the medical support provided by the army camp medical officers. My stay at Bandarban was short, as I returned to Mirpur to attend the Staff College. My well-wishers knew that I was with the field unit for short t ime so they engaged me in a number of long range patrols or counter insurgency operations. In hindsight I can say this exposure benefitted me and helped me to improve my professional ability.

Student Staff College—1983

I attended Mirpur Staff College as a student in 1983 when it was still in its formative years where the college could provide minimal logistic support. I attended the 8th Army Staff Course, and the Army Wing had only around thirty students, plus one Air Wing Syndicate and one Naval Wing Syndicate. Together the staff courses included around fifty students including a few from overseas. Computers were non-existent in the Bangladesh military in the 1980s, and all our assignments had to be handwritten when they were submitted, including the script of the Commandant's

Paper. PowerPoint, which has made presentations so easy nowadays, was unavailable in 1983. At the time, I had heard of presentation software called 'Harvard Graphics,' but that required a huge multi-lens projector: all that is history; probably one can find that projector in the museum now. Our classroom presentations were mostly on handwritten 'View Foils' using an 'Over Head Projector,' which may not even exist in museums now. BMAT Team had introduced some 16mm training films, which came on loan from the UK Ministry of Defense. These projectors have also gone out of use now, as video clips became more popular and easier to use.

Our last joint services exercise was based on the setting of an amphibious landing south of Cox Bazar. On this tactical exercise, I had the appointment of 'Combined Amphibious Task Force Commander—CATF Commander.' During the preparation week, I, along with my naval component commander and air component commander, kept discussing the options and plans, but we needed more time to come to a conclusive plan. As the team leader, I was required to present the plan, and my team members were worried; now what? Over the years, I had become confident in giving oral presentations. I was not so worried because whatever we had discussed throughout the week was all in my head; it was just that the script was not written. I quickly prepared an operational map with markings showing everything that one would expect to see in such a tri-service operation.

On the presentation podium with me, I had a few pages with notes and a few blank pages. I told the officer assisting me that he must appear confident and point to the locations on the map as I did the presentation. My impromptu presentation, done without a written script, was a great success; the senior Navy Instructor, Captain Ashby, a British Navy Officer who was taking in the presentation, was very pleased and left me with lots of words of appreciation. The officer assisting me on the map later told me that he was worried I would get into trouble, since I did not have the full script in hand and was surprised to see me giving a presentation holding blank pages. The year at Staff College was full of many other sweet memories, including my frequent motorcycle rides to the city and other places on my Honda 110 motorcycle. To buy this motorcycle, I had taken a car loan from the army, which at the time (1978) was Taka 15,000/- and the motorcycle cost me, Taka, 13,500/-. We lived in a time of great financial constraint. Military officers strove hard to attend staff college because this is considered an essential step in enhancing one's military career.

5 East Bengal—Return to the Parent Unit

Upon completing my staff college, I was promoted to the rank of Lieutenant Colonel and was posted as Commanding Officer (CO) of 5 East Bengal Regiment. I had served in 5 East Bengal as a young officer from 1974—1976 in Comilla. I considered myself very fortunate to be given command of my parent unit, which is a great pride for infantry officers. I assumed command of the unit in Jan 1984 in Chittagong, just before it moved to Dhaka Brigade. The month of January is not a good time for military families in Bangladesh, because officers are away with the units for weeks together for the annual field exercises. I arrived in Dhaka with my unit 5 East Bengal and had to move out for the annual field exercises immediately. While I was in the exercise area in Narshindi, I was directed by Army HQ to go to Infantry School Sylhet for a few days to review the officer's weapons course syllabus. With all the travel army wanted me to do I was concerned about my wife because she was expected to deliver our second child at any moment, and I had to find somebody to drive her to the hospital when the labor pains would start. Fortunately, my mother-in-law came to my home to look after her. Even today, I am extremely grateful to my military driver, Manzoor, whom I had left behind and telling him to rush to my house the moment the telephone rang. At my residence in 1985, I did not have a mobile phone or a regular landline; the only thing we had was an army field telephone which had a one-to-one wired connection from my house to the unit car park. One had to crank the phone handle a few times to make a call. I left for Sylhet at night by train, and in the morning, when I reached Sylhet railway station, I received the news that my wife was taken to the hospital in the early hours in the morning by my military driver and she was blessed with our second son, Ishtiaq, that morning. I am deeply indebted to my driver Manzur, who jumped out of his bed at 4 am and rushed to my house to take my wife to the hospital. I commanded 5 East Bengal Regiment for two years in Dhaka January 1984—January 1986. During my time with 5 East Bengal, I focused on training the weapon crews, especially the battalion mortars, because I was a weapons instructor at Infantry School.

Instructor Staff College & The Early Days of Computers

Immediately after completing my two-year unit command tenure, I was posted to staff college on a teaching assignment or Directing Staff. I taught at the Staff College for two years, 1986-88. That was the time when most of the British instructors had left, and there was tremendous pressure on us to re-write the precis issued to students and exercise papers, the purpose being to align the teaching at Staff College to the Bangladesh environment. One of the classic examples of the BMAT time exercise that required change was 'Agni Bina', which involved covering troop battle by an armored brigade, which is unrealistic on Bangladesh riverine terrain. The setting appeared to be a typical battle for the British Army in Germany, with the armored brigade fighting a withdrawal battle faced with a massive Soviet Army onslaught. This CPX (*Command Post Exercise*) was run on the Jessore—Benapole road for the first few courses. To run this exercise in Jessore required huge logistical support including airlifting all students and staff by air force transport aircraft. Later we moved this CPX to Joydepur—Mymensingh Road and we changed the exercise settings. In those early years, Staff College use to have a separate one week package each for the Joint Warfare Team from UK, the Department of Air Warfare Team from UK, the NBC (*Nuclear Biological & Chemical Warfare*) Team from UK and the TRADOC (*Training and Doctrine Command*) Team from USA. In my view, these packages were very useful because the team's presentations gave the students a flavor of high-intensity warfare, electronic warfare, and the use of advanced weapons systems.

Staff College Mirpur introduced computers in 1980, when very few people in the military knew how to use them. Computers were not accepted as an essential office tool, especially by the Bangladesh military. I was among the few odd people who spent time learning about them. Because of this, most people viewed me with suspicion, assuming I was wasting time and playing games, which was a common negative perception about computers at the time. Computer users today may find it difficult to believe that the computers we used at the time had only 640K RAM, yes 640K RAM memory which was used for the computer to boot on DOS only. Storage on the computer itself was not possible. Monitors had a dark screen with green-lighted writings like the old dump terminals. We had to remember the commands that we typed on the command line for every task, even for simple things like saving files. The OS on our computers was CPM 86 from ICL computers UK. Windows which is such a popular OS now, did not exist at the time, at least not in Bangladesh. To start the computer, we were required to insert a $5^{1/4}$ inch floppy disc which had the OS software and then push the power switch. After the computer booted with the cursor blinking, we would remove the OS floppy disc and insert another $5^{1/4}$ inch floppy disc with the application software. We used a software called' Word Star' for word processing at the time. A few years later, Word Star was replaced with another improved word processor called 'Word Perfect', and now we have the 3rd generation word processor 'MS Word'. Similarly, for the spreadsheet, we were using Lotus 123. With the arrival of 'Windows' software, computers became friendly and gained popularity.

Author sitting in front row, extreme right, along with other teaching staff at Staff College Mirpur, 1987

Over the next few years, I became an expert on word processors and spreadsheets. It would be difficult for people to believe now that email was non-existent in the 1980's, at least in Bangladesh. My interest in computers increased, and on my next placement as Military Secretary to the Army Chief, I was able to push in one of the early models of Macintosh box-like computers with 1MB RAM. This was my first experience with the graphical interface screen clicking icons, and we were super excited that we did not have to type commands. This was about when I also got my hands on a Toshiba laptop in 1988 when very few people had seen laptops in Bangladesh. It took quite a while for offices to introduce computers specially with government offices. The best we had at the time was IBM Electric Typewriters.

Secretary to Army Chief

After serving at the staff college for two years, Jan 1986—Jan 1988, I was transferred to Army Headquarters as Secretary to Army Chief. It's customary in the army for the Army Chief to pick officers on his staff amongst officers who have worked with him and whom he knows personally. So, it was a surprise for me to have been posted as the Military Secretary to Army Chief, Lieutenant General Atiqur Rahman. I had never met him and had never served anywhere around him. I was told later by the Military Secretary that the Army Chief's opinion was that it is the job of the Military Secretary to find an appropriate person, so the Military Secretary picked me up for the job. A similar incident occurred a year later when the Army Chief's ADC (*aide-de-camp, an officer on personal staff to assist and accompany the general officer*) had to move out for a different assignment. The Military Secretary went to see the Army Chief with a panel of likely candidates and requested the Army Chief tick off the officer he wants to be his ADC. The Military Secretary was taken aback when the Army Chief told him that it was the job of the Military Secretary and not the Army Chief. I served as the Secretary to Army Chief for nearly three years until his retirement. During my tenure as the Military Secretary, I was not required to carry files into his office, except for the file on which he had put a note for me to see him. I would quickly glance through the files more to stay in the picture of what files are going in and what files are out. All the files which were sent to the secretariat would be quickly cleared from my office and sent to the Chief's Office on the same day. Army Chief, General Atiq was very quick in disposing of the incoming files. Most of the files would be cleared the same day before he left the office. As a standard practice, files that he could not clear during office hours would be put at the back of his car to take home. He would see the files at night and return them the next morning. Another interesting habit of General Atiq was leaving the office a few minutes after 2 pm every single day. This also gave us the opportunity to leave the office immediately after his departure. However, one day he was in his office well past 2 pm, so I walked into his office to check. He smiled and looked at his watch only to realize his watch was showing the wrong timings. That's the only day in three years he was late in leaving office. General Atiq would usually only encourage officers to seek courtesy meetings with him during office time if the person had an official agenda to discuss. Organizing and coordinating the Army Chief's visit was one of my important tasks. A big departure from the past was for the hosting organization to not put out lavish dinners or lunches and to keep the event simple. During a visit to the Bangladesh Military Academy, the academy commandant was very upset in the morning because the Army Chief, General Atiq only asked for a cup of tea and a slice of bread. The large number of items prepared for the breakfast were left untouched; I guess the chef must have been unhappy too.

The other important task left for me was to organize and coordinate the training visits especially during winter exercises when the military units were engaged in tactical maneuvers. I would normally read through the exercise papers sent by the concerned military unit and write a short brief for the Army Chief to read through before the visit. The purpose of my brief was twofold: first, for the Army Chief to leave the impression that he has indeed read the exercise paper that has been written so painstakingly by his hosts, next, for the Army Chief to be asking questions relevant to the tactical settings of the exercise. My brief would also include a list of queries that the Army Chief could raise and notes on the reply to expect. The Army Chief I served had one great quality, he would quietly listen to staff opinions which would not always be to what would please him. Many a time, I could tell him things that I felt would be incorrect for him to do, and he accepted such suggestions in a positive spirit. I felt the high position of the Army Chief insulates him from the real world, and therefore, I would take the liberty to tell him what other people would be thinking and their expectations on different arising issues.

Counter-Insurgency Operations, Chittagong Hill Tracts, Diginala

After the retirement of the Army Chief, I was posted as Commanding Officer of 6 East Bengal, located at Diginala Chittagong Hill Tracts. This was my second unit command. The area north of Diginala, especially Babu-Chara and further north, was infested with insurgents and their sympathizers. So, on a routine basis, we were frequently committed on counter-insurgency operations. Any movement north of the Babu-Chara camp required careful planning, considering the high likelihood of confrontation with armed insurgents. In the area, we had a small base named Jarulchari, which was about three hour's walk through the jungles north of BabuChara Bazar.

To dominate the hill jungle area further north up to the border with India, the Brigade HQ decided to establish an additional camp further north. The area was a kind of 'RED' zone where we encountered hostile fire whenever our patrols ventured into the area. The standard procedure for helicopter pilots was to fly at a height out of the hostile fire. I had also noted this flying procedure earlier; helicopters would fly low before landing after security forces patrols secured the area. We had selected a few places on the map as likely camp locations based on our information on insurgent movements in the area, the location of tribal villages, the proximity of other security forces camps and access to river water. Access to secured water points becomes a critical need to meet daily cooking and washing requirements. High heat and humid conditions cause sweating and exhaustion. So, a dip in cool river water is refreshing; therefore, the water point needs to be secured before use. We placed a few armed soldiers or a standing patrol to watch the area for insurgent movements. For the selection of the camp location, our first priority was to judge how vulnerable the place was to insurgent fire from neighboring hills. So, we looked around to confirm there were no dominating grounds from which insurgents could threaten the camp by rifle fire. This had to be balanced with the survival need to access a secured water point. We also learnt that drinking water could be found when fresh bamboo is split open, so having bamboo bushes close by was helpful. Bamboo water is probably the purest water, but it has no taste. To select the campsite, we went for a helicopter reconnaissance flight, and the Brigade Commander showed me a few possible sites. The helicopter circled around to allow us to see as much as we could before returning to base. This location was about four hours walk through the hill jungle north of our nearest camp. The area was almost devoid of human habitation. This was my first experience in establishing a new camp for counter-insurgency operations deep in the insurgent-infested area.

Based on the site shown to me from the helicopter from a height of about three thousand feet, about two weeks later, I started trekking for the site accompanied by two strong fighting patrols and a human train of tribal porters. Porters were employed only for some special occasions to carry the essential loads. We estimated that for one week we were not likely to get the helicopter resupply and therefore used porters to carry everything we needed for the week to set up the camp. Security protocol in practice required us to prepare the helicopter landing pad and to secure the area around it. We also had to ensure a clear approach route so that pilot could see the marked landing pad from a fairly long distance and would have a clear exit route with no intervening hills in the immediate vicinity. Further, as a standard security procedure, before any landing, few patrols were sent out for area domination and to prevent insurgents from taking shots at the helicopter when it is most vulnerable during landing approach and departure.

We started early in the morning and reached the new camp destination about midday. After deploying the soldiers for perimeter security, I walked around to confirm my selection for the final site. I found a nice covered approach to the possible camp next to the water stream. The presence of a water source close to the camp was good news, as it just makes life so much more comfortable.

I planned to spread the camp around the hilltop with nothing on the skyline. We situated ourselves a little below the crest line on the slopes. Being on the hill slope only exposed us to hostile fire from one direction. Our camps in the hill tracts were essentially a cluster of huts made of bamboo and sun grass. For safety, I instructed that inside the bamboo huts, the sleeping bunks should be dug down a few feet below the ground level so that soldiers, while asleep, were protected by mother earth. Digging down was done by my soldiers who were off sentry duty. We required a lot of bamboo and straws to build the living huts and other camp facilities.

Here comes the additional purpose of bringing the tribal porters. These porters were very good at using the locally available material from the jungle to make huts. We used their skills for construction purposes paying in cash at local labor rates. The downside of keeping the tribal labor was that they had to be provided food from our very meagre food stock. In about three days' time, I allowed our porters to go back primarily to save on food. By this time, we had earned the confidence of tribal people from the locality, to come and work for us on cash payment. There were no tribal villages in the area, but only a few scattered families, so we could engage only a limited number of local workers. For security reasons, the locals worked only when armed guards were posted around them and were checked before entering the camp area. It was safe to assume locals would be insurgent sympathizers. For about a week, we slept the night in the bushes under the open sky as we worked to build the camp facilities. Luckily, I had carried my personal sleeping bag, which I had bought in USA a long time back, and this sleeping bag came to good use.

It was about fifteen days since we arrived at the new campsite when our new camp was put on the visit plan of the Army Chief. I thought that was pretty unfair for the little time I had spent on the location for camp build, and I felt that I was being subjected to further failure tests. We prepared for the visit to every possible detail, including preparation of the helicopter landing pad and practiced a few times sending out security patrols to cover the dominating grounds around. I was very pleased to see that the visit went smoothly and, in my judgement, the Army Chief, who knew me well otherwise, went back happily. I returned to home base, my battalion HQ leaving the further development to the new camp commander. This camp was named '*Dhanpata Chara*' camp, Dhanpata being the name of the small stream nearby. I visited the camp a month later, but this time on a helicopter on a monthly logistic support mission.

For counter-insurgency operations, on a routine basis, we had two major operational activities. Every day in the morning we would send out 'Route Protection Patrols' to clear and secure the major roads and traffic routes in the area that civilian and military vehicles used. The procedure for route clearance was for our patrols to physically walk along the road to ensure no ambushes were in waiting and no roadside IEDs had been placed by the insurgents overnight. We called this the '*Route Protection*' drill, which had to be done daily. Route protection patrols would hold temporary positions along the checked road during usage hours and withdraw by late afternoon. Both military and civilian vehicular traffic were barred from using the road until Route protection patrols were in place the next day. The next important work for us was intelligence gathering on insurgent movements based on which we planned raids and ambushes. A lot of resources were committed to pacification programs like the support for primary schools, agricultural projects, and road building. The purpose was to '*win over hearts and minds of the people.*' To us, this was an essential requirement for a successful counter-insurgency campaign.

To reduce the insurgent threat and influence in the area, I worked on a strategy to subject the insurgent movement to the same vulnerability of surprise ambush that insurgents typically present to security forces. So, I encouraged my camp commanders to practice sending out small patrols overnight a short distance from the camp to suspected insurgent movement tracks like foot tracks leading into or away from local villages. My directive was based on my belief that if we succeed in even a single ambush, we will make the insurgents feel unsafe in the area.

I had planned and executed one such ambush on a bigger scale. I sent a strong fighting patrol with around thirty soldiers and two officers to lay an ambush on a suspected insurgent movement track waiting for two nights. The patrol slipped out of our camp late at night so that locals do not see the movement and they headed for a hill range called 'Alamgir Tilla—Alamgir Hill', named so because a long time back, a soldier named *'Alamgir'* died on that hill due to insurgent fire. This ridge line served as an operational boundary between two neighboring operating regions. Boundaries are comparatively less secure areas because of the belief that the other side is taking care of, and that may not be true.

For this operation, the ambush commander was Major Mahbub, a very courageous officer on whom I could rely upon for the most challenging and dangerous missions.

They reached their spot at night and laid out the ambush in waiting. My instructions barred all daytime movements. The first twenty-four hours passed without any contact with insurgents or incidents. It was on the second night at about 3 am I received a radio message that Maj Mahbub had received a bullet injury in the stomach. It transpired that one of the soldiers next to Mahbub misfired his rifle, and the bullet hit Mahbub. To my relief, I was told the first-aid field bandage was wrapped around the wound, and Mahbub seemed to be stable as he was talking normally. I instructed them to withdraw quickly to a nearby camp about three hours away. The patrol made an improvised stretcher with bamboo and started the withdrawal. Simultaneously the nearby camp was instructed to send out a link patrol to secure the route and to link with Mahbub's patrol. The recovery operation was successfully completed with the injured officer arriving at the camp. I did not want to wait for the helicopter flight which would not come until the morning. So, I decided to send Mahbub by ambulance to Army Hospital at Khagrachari, the same night over the unsecured road. This was a difficult decision for me, to evacuate the patient or wait for the road to be secured the next morning. At Khagrachari, the army doctors were in waiting; they quickly operated, removing the bullet. Mahbub recovered and returned in a week's time. To me, we were executing a very interesting 'opportunity ambush' operation, which had to be aborted on the second night due to the accidental fire incident.

I wish to mention another insurgent operation I undertook personally during my short tenure in the Chittagong Hill Tracts. One day we received location information of one insurgent toll collection team.

Insurgents did not maintain permanent camps inside Bangladesh, for the insurgents, it was always a cross-border operation. After crossing into Bangladesh, insurgents roamed around in their area for a few days and would then go to the safety of their bases across the border. Local tribal population supported willingly or under coercion. Larger armed insurgent groups or fighting insurgent groups generally stayed along the border belt where they could quickly exfiltrate for safety. Insurgent toll collection party comprised of two or three men only, and for security, they changed their location every night. The insurgent toll collection party also functioned as the political propaganda arm for the insurgents. These toll parties maintained close contact with the local tribal population from where they could also collect information about the military and other security forces.

We generally employed our trusted human contacts from the villages for intelligence on insurgent movements. Informers worked either to get a financial reward or simply out of vengeance against the insurgents who may have caused them harm in the past. People who have been harmed by the insurgents in the past become good informers. Timely intelligence is crucial to catch armed toll collection parties because they stay in the area for a few days only. Furthermore, the location information would generally be good for one night only because for survival insurgents change their locations every night.

It was around mid-day when we received information on the possible site of the insurgent toll collection party. In discussion with the Brigade HQ, it was decided to immediately send out a fighting patrol the same night to raid the place. Usually, I would stay by myself at Diginala

Battalion HQ Camp and my wife, along with our two small sons, would stay in Chittagong city, some five hours drive away. My wife would normally visit me when schools were closed. On that day, my wife had arrived with my sons for a short visit, and my dilemma was how to tell my wife I would be out that night on a raid mission deep into the reserve forest. As a coverup story, I told my wife that I had a prescheduled visit plan to a close by camp where I was required to stay the night and would return the next morning. I was relieved to see my wife as very normal, for she did not ask me any probing questions.

Patrol assembly and preparation were done after sunset so that insurgent spotters who might be observing the camp from a distant hill did not see troops assembly. In the insurgent-infested areas, local sympathizers generally acted as their informers. They went around their routine work but keep an eye on the military. Information on army troop assembly and patrols leaving the camp is immediately passed on. The value of such information is time-based. Timely information helps the small insurgent groups to move away to safety. If the insurgent fighting element is in the area, they can plan an ambush on the military patrol. That night I led the fighting patrol and walked the whole night in drizzling rain. We were walking through dense jungles and did not have GPS or any other navigation aid. We relied on well trusted local guides. These people had used these tracks for years and could recognize the jungle paths even on dark nights. For security reasons, our wireless was kept silent unless we faced an emergency or until we encountered the insurgents.

Around midnight, we came across a narrow stream with a strong water current caused by heavy rains upstream.

In the hills, after rains even small river streams can turn into rapids for a few hours making the crossing dangerous. Our local tribal guide told us to wait for an hour or so for the water current to slow down. At the river side we waited for about an hour, but the river flow did not slow down sufficiently. I was getting impatient as time was passing and we had to hit the target before sunrise. With the night hours slipping away, I decided to take the risk and attempt to cross the river. Water was still about waist high, I instructed the patrol to form a human chain by holding on tightly to the waist belt of the soldier in front. Together we created the chain, providing stability to each other. This worked very well, and we were able to cross the river. We continued walking through the night and just about first light we reached the target, a Jhum Ghar, (*an elevated temporary straw and bamboo shelter made by tribal farmers to watch over their farmed crop area*).

Jhum Ghar is a lonely hut away from the village in the cropped area. During the cropping season, tribals stay in Jhum Ghar to scare away animals like wild boars and elephants that damage the crops. As we came close to our target Jhum Ghar, the insurgents sensed the presence of the army patrol and fired on us. We returned fire and rushed onto the target. It had rained through the night, so the slopes of the hill were muddy and slippery. Many, including me, slipped, and rolled down the hill a few times as we rushed uphill. The charge was led by Captain Maksud, who retired many years later as a Major General. I also had Captain Jamal in the patrol who as Lieutenant Colonel died of a heart attack a few years later at a young age. Both Captain Maksud and Captain Jamal were very courageous officers and would never hesitate to lead fighting patrols to ambush or raid insurgent targets. On this raid we captured some toll collection receipts and one 9mm Sten Gun. Personally, I am very proud that this was the first weapon recovery for 6 East Bengal and happened with me leading the patrol.

At the time, I was in my mid-forties and on my second unit command. I have to admit that physically, I was stretched to my limit walking the whole night on this long-range fighting patrol. I guess what kept me going was something that was deeply ingrained in one's minds during training at the military academy. We were brainwashed with the Urdu language word 'gairat' an army jargon meaning self-respect which implies you die or do the job but cannot give up. A deeply ingrained teaching that drives the fighting soldiers. We had completed the mission successfully and were very tired having walked the whole night.

The more difficult task now was to get the patrol safely back to the home base a few kilometers away. Security forces suffer most causalities while returning to home base because soldiers tend to let down their guard and rush back. To avoid being ambushed, I planned our walk back on a totally different route, and we remained very cautious until we reached the outskirts of our security perimeter. News on the successful raid and weapon recovery had already been passed on radio, so a small group of jubilant officers and soldiers came forward of the base camp to welcome us. By the end of 1991, 6 East Bengal had completed its three-year tenure and moved out of Chittagong Hill Tracts to Jessore. The unit was earmarked for UN Mission to Kuwait, and the unit started preparing accordingly. In the meantime, I got my long-due promotion to the rank of colonel and was sent as Senior Instructor Army Wing at the Staff College.

UK Training Visit

By 1991, Staff College Mirpur was in its final phase of transition as the British Military instructors had departed and Bangladesh military officers had fully taken over their responsibilities. Bangladesh officers assigned to work as Senior Instructors at Navy Wing and the Air Wing had completed their one-week orientation visit to the UK Staff Colleges and were now fully functional. Initially, Army Wing did not have a senior instructor, it came as an additional responsibility for the Chief Instructor to look after the Army Wing. The new position of the Senior Instructor Army was created before the departure of the last British Chief Instructor, Brig Wolverson. The first Bangladeshi officer posted as Senior Instructor Army left for brigade command in three months. I joined as the Senior Instructor Army Wing a few months later. To start, I had seven or eight syndicates under me, which were later split into Division A & Division B with separate senior instructors. A few months after I joined, I was deputed for an orientation visit to UK Staff Colleges. Earlier in my military career, I spent a lot of time as an instructor, as a Captain (*weapons instructor at COMBAS Infantry wing*), as Major (*tactics instructor at COMBAS tactics wing*)) and as Lieutenant Colonel at Staff College (*DS—directing staff*). So during my visit to the UK Staff Colleges, I was very keen to visit a few training institutions in the UK. The Defense Attaché UK in Dhaka was coordinating my orientation visit, and he was kind to accommodate my request. He managed to get from the UK Ministry of Defense one extra week approved for visits to different training institutions and a few days of attachment with a field unit in training. The trip included a field trip with the para brigade at Salisbury and Norfolk.

In 1993, the UK still had three separate staff colleges for Army, Navy and Air Force. In the UK, they had just started looking at the possibility of establishing a single combined staff college like they had helped us to create in Bangladesh. However, I found that the army, navy and the air force resisted the new concept, as the they prioritized their own service heritage and history and wanted to preserve the individual service staff college. After years of effort, it was in 1997 UK was finally able to set aside the interest of the service and set up the new Joint Services Command and Staff College. During my visit in 1993, I first stopped in and visited the Naval Staff College at Greenwich. I was very fortunate to have seen the narrow tunnel-like corridor at the old hospital through which during the World War II the wounded soldiers arriving on ships from overseas were taken into the hospital. I was also very honored for my lunch hosted at the 'Painted Hall', a UK national heritage. The vast ceiling with the age-old paintings was something like one would see at the Vatican, leaving the onlookers in awe. The next day I travelled to the UK Air Force Staff College at Bracknell. It was just a day's visit and I returned to London by train in the evening.

The last leg of my UK Staff Colleges visit was at the Army Staff College at Camberley, where I spent maybe three days. I was accommodated at the Camberley Staff

College officer's mess. I was indeed fortunate to have the company of a Lieutenant Colonel from amongst the teaching staff assigned as my conducting officer. He gave me a good walk-through of the course contents, course conduct, exercises being run and problems they faced. He accompanied me every evening for dinner. On the day of my departure, I asked for the bills including my bar bill which was appropriate for me to pay. Guess what, I was told I was being treated as an official guest, so the staff college was paying for all my bills. I asked what about the bar bill, it was very nice to hear that was also covered, as well. This was a pleasant surprise for me because I had heard from senior officers who had visited UK earlier that they paid all their bills and were never provided any official transport. Not the kind of hospitality that we extend to visiting officials in Bangladesh. My case was totally different, I guess my former boss, Brigadier Wolverson, former Chief Instructor Mirpur Staff College, had coordinated my visit in line with the hospitality that he enjoyed in Bangladesh. Like all other BMAT members he must have felt an obligation to look after officers from the Bangladesh military. They deeply appreciated the hospitality they had received over long years of stays in Bangladesh.

Of more interest to me was the week after, which I spent visiting various training institutions. I started my visit at the Armor School Bovington where I was able to get into the Challenger Tank simulator and later in the day got an opportunity to fire the Scimitar at the firing range. I was also given the opportunity to drive the new Scorpion armored reconnaissance vehicle. I was very impressed with the two NCO's (*non-commissioned officers*) assigned for my field part of the Bovington visit; both were very smart, knowledgeable, and very professional. I also able to visit the tank museum at Bovington and see the very first tank that appeared during WW1. The most interesting experience of my Bovington visit was at the armor school officer mess where I was staying. In the morning at the school, I was told to stand in front of the school flag for an official photograph. I was moved to notice that they had raised the Bangladesh Flag next to the school flag to honor my visit, indeed very kind of my hosts and a very rare gesture. Next, I went on a half day visit to the School of Infantry where the I found the training of snipers most interesting. They were training on 0.50 Caliber heavy rifle with folding bipod. I was told these snipers operated in teams of two men behind enemy lines and were trained to take pot shots at a distance of 1,000+ Meters. These snipers were being trained to take out enemy commanders from long distances. Many years later during a peace keeping mission, I saw the extensive use of snipers during the Yugoslav war in Bosnia. As I write now, I hear about snipers killing Russian senior officers from long distances in Ukraine. Snipers with 0.50 Caliber heavy rifle can also effectively take out soft or lightly armored vehicles. Equivalent to 0.50 Caliber would be 12.7mm in Bangladesh Army. I guess a heavy bullet provides ballistic stability over long distances. Later in the day I stopped at Infantry School Support Weapons Wing where I was shown training on French Milan ATGM and use of Thermal Sights. This was my first experience with looking through the thermal sight even in day light I could see a horse behind a bush at a fairly large distance, because the horse body heat was much more than the surroundings.

What they were trying to tell me was that a tank hiding behind a bush could also be seen even during daytime using thermal sights, because the tank's metal body will always radiate more heat than its surroundings.

I was lucky to have a dedicated car for my travel like a VIP, a rare practice in UK for visitors of my level, army colonel. After lunch I left for the Special Forces (SAS) School at Beacon Brecon in mid-Wales. I reached the school in the evening. The school commandant, Lieutenant Colonel, was kind to host me and I stayed at his house. I remember the school commandant telling me Beacon Brecon was chosen as training base for the special forces because it had very rugged mountainous terrain as well as extremely cold and windy weather conditions. In this harsh environment, only the fittest of the fit can survive when left in the open. The next morning, I had a glimpse of Infantry Squad Commanders (*infantry section commanders*) training of the Territorial Force. In the Bangladesh Army, undertaking fire and move drills firing live ammunition was unthinkable at least in my time. Here I was walking behind a Squad, equivalent to an infantry platoon section, practicing fire and move live fire. The squad was divided into two fire groups and as they attacked and approached the objective the enemy fire was simulated. The squad advanced with one fire group providing covering fire to the other group as they leap frogged ahead. This was a simulated practice of an infantry platoon battle drill. They also practiced tossing live grenades inside a marked MG Bunker, obviously with no human crews inside the bunker. Of particular interest to me was that these were territorial force soldiers, civilians who had come for short refresher training, not the regular army soldiers.

My next destination objective was to witness the British army training at a tactical and unit level. My visit started with HQ Para Brigade at Salisbury, where I was briefed by the Brigade Major and later had a short meeting with the Brigade Commander. The British Army was just starting to practice the American Army 'Intelligence Preparation of the Battlefield—IPB.' I noticed that Brits loved their old tactical appreciation over IPB use as much as we in Bangladesh Army preferred the old-style tactical appreciation format, at least until my time. This was a time of uncertainty for the British Army officers. The UK Government had decided to reduce the size of the army and every day officers were receiving their letter of retirement. The Brigade Major told me that the day before, a few officers of the brigade had received such retirement letters and were sad to be leaving the army, especially the para's.

After the Brigade HQ, I was driven to 3 Para Battalion where I spent a long time with the battalion 2IC (*second in command*) who briefed me on how they planned the battalions training programs. It was interesting for me to compare every part of their training with what we did in the Bangladesh Army at the battalion level. The process starts with the year's Training Instructions issued by the Brigade Commander. The Brigade Commander provides a list of operational capabilities that the battalions have to be prepared for by end of the training year. For example, the brigade commander wanted each battalion to have two 'path finder teams for night insertion,' and one company to be ready for tactical low level night para drop. Based on such requirements, the Battalion Commander issued his training instructions where he would mention the operational capabilities that each company must achieve within the year. Compared to Bangladesh Army Training instructions which tend to be big, bulky documents, the idea being the thicker the better, the training instructions that I was shown here were simple and concise, just a few pages of documents. I did hand over all my document collections from the UK visit to Army HQ Military Training Directorate as reference samples. I always thought we were misdirected in producing thick bound booklets as training instructions. Our training instructions documents amounted to micro-management like producing lists of training classes to be run and even mentioning the number of people to be trained. Since I was visiting a para unit, I remembered the instructions from the Brigade Commander to Battalion commander for low level night drop capability and 'path finder teams' that the battalion must prepare. The Battalion Commander subsequently assigned the task responsibility to a specific company and the company commanders then did self-assessment as to what kind of training they needed for the company. So, each company ended up doing different forms of training. Micro level planning and execution was being done by the company commanders. Company commanders spelled out their detailed training plan to the 2IC (*second in command*) along with the support they need such as transport, firing range, and allocation of air resources for para drop practice. Looking at what the battalion commanders did in Bangladesh, I asked the 2IC where does the battalion commander fit in? His reply was the commanding officer has very little to do at the micro execution level, he is the King. With the three battalions I commanded I wished I was treated like a King with a 2IC running the whole show.

Later that afternoon, I was taken to RAF Base Lyneham. I along with some soldiers boarded a C-130 which flew us to an abandoned airbase at Kings Lynn, Norfolk, a small city north of London on the east coast. I was told this airbase was earlier used by USAF but now lay abandoned. 2Para battalion was camped at Kings Lynn, Norfolk, at the airfield for operational training exercises. I had mess-all lunch with 2Para soldiers and was briefed that they would be practicing airfield seizure operations later in the evening, a likely task for 2 Para in future if British forces were to be inserted in Bosnia—1993 scenario. Conflict in Bosnia was brewing up and at that time I did not even dream that a year later I would myself be in Bosnia as UN Peacekeeper along with the UK forces. Incidentally the Major General Mike Rose who had been the commandant of the UK Staff College Camberley during my visit to UK in 1993, was the UNPROFOR Commander in Bosnia Herzegovina when I was assigned to that mission.

I had the afternoon free and my conducting officer, a Captain, asked me if I wanted to see any other facilities. I was told I could visit FIBUA (*Fighting in Built Up Area*) village mock-up training facility, or I could visit a nearby field firing range. I opted to the visit the field firing range. I was impressed to see the Captain making a call on his cell phone, and in a short time a Scout Helicopter arrived. We flew to the field firing range. As we were about to land there we noticed a car going out of the facility. The Captain identified the person driving the car as the Range Officer we intended to meet. The helicopter circled low around the car and with hand signals we indicated we intended to visit the firing range. The car went back to the range office area where we landed.

With UK Para Brigade at Kings Lynn Norfolk, 1993. Author on left (Source: Author's own records)

The Range Officer, a retired Major, did not complain that he had to turn around, rather he welcomed me and took us to his office where he briefed me on range organization and management practices. The range was being operated by this retired Major with a few locally employed people. The range had reasonably comfortable accommodations for the user units with kitchen and allied facilities. An interesting thing I noted was that the army did not dispose or clear the unexploded shells or bombs. I guess over the years dozens of unexploded shells and other ordnance must have piled up inside the impact area. Therefore, for safety concerns locals were barred from going inside the impact area. During my days as a weapons instructor, I used to take the infantry school students at least two times a year to Chittagong Hathazari field firing range for live firing of heavy weapons. On the range area, we had to carry whatever we needed and set up our camp. Forty years on, I understand things are much better now and range now has accommodations and kitchen facilities.

My interest at the time was to study the range organization and how the UK Army allocated slots to a large number of units. I was told, whatever be the schedule, if a unit gets an operational deployment order, they get priority over every other unit.

With UK Para Brigade at Kings Lynn Norfolk, 1993. Author inside C-130 taking off ramp still open after troop insertion on hostile airfield seizure simulation (Source: Author's own records)

In the evening the para battalion exercise was about to start. For the exercise, I got on onboard the C-130, the elongated J Model, which had one platoon from 2 Para Battalion and one Scimitar recce vehicle loaded. That evening they practiced two forms of the tactical exercise, the first was to practice the seizure of an airport. For the first exercise we took off and after a circuit, the aircraft approached the airfield which was being simulated as the enemy airfield. The aircraft made a short and very steep approach, more like diving into the airfield, and then landed. The aircraft continued moving slowly as the ramp door opened, keeping a little above the concrete runway to avoid damage. While the aircraft was rolling slowly, the para troopers rushed out of the moving aircraft and the Scimitar recce vehicle also exited on the runway with a hard landing. The aircraft got airborne the moment the last of the troopers exited. The aircraft exposure time on the hostile runway was very small. Later the same evening, I joined on the next exercise which was for one 2 Para company to execute a tactical airdrop from very low altitude, I was told the drop height was from 500 ft. As a cadet in the military academy, I did my basic jumps from C-130 aircraft from altitude of 1200 ft. Low altitude drops can be done by highly skilled soldiers only, and I was witnessing some of the best trained paratroopers who were otherwise considered as elite forces. After the drop the aircraft flew back to RAF Base Lyneham and I returned to the officer's mess for the night sleep. My visit to the para brigade was over. For me the day was very exciting and busy gaining a lot of real time experience and knowledge.

Next, I had a long drive to Exeter, a city located on southwest costal region of England. I was visiting the Brigade HQ of the Territorial Brigade at Exeter. The Exeter Brigade Commander who was hosting me was Brigadier Wolverson. My boss as Chief Instructor at the Staff College in Bangladesh. He made arrangements for me to stay the night at his house and had his car and his driver at my service. Because Brigadier Wolverson had stayed in Bangladesh for three, long years, I guess he wanted to return a part of the hospitality he had enjoyed in Dhaka. His wife even insisted on doing the laundry for me, besides making the meals, and other arrangements she painstakingly put into place to make my stay comfortable. When I tried to do the laundry, she laughingly told me, Salim I have lived in Bangladesh, and I know you would never do that by yourself, so let me do it for you. I was enjoying the benefits of Staff College Mirpur affiliation.

Staff College Mirpur used to get training teams from the UK such as the Joint Warfare (JW) Team to conduct the JW package. Brigadier Wolverson arranged for me to visit the Joint Warfare School at Poole. I went around the school facilities, including taking the opportunity to see the UK Marine Commandos under training. With the visits over, I had to return to London from Exeter. Brigadier Wolverson arranged a car with one of his soldiers as driver to take me to Portsmouth where I was invited to stay a night with my former staff college colleague, Captain Peter, Senior Instructor Navy, now retired. Peter was kind to take me around to show the naval heritage ships, later in the day I returned to London. So my UK trip went very well with all transport support from UK Ministry of Defense and my Staff College ex-boss Brigadier Wolverson. I was grateful to him for coordinating my visit down to every detail, and specially taking care of my logistical needs.

On my return from the UK, I continued serving at the staff college as Senior Instructor Army Wing. Work at the Staff College went like clockwork, so one would never be under a sudden surge of work. Staff college was a good place for family life. Beginning in1994, I moved

from Staff College to become the Officiating Commander
of Rangpur Brigade.

**Author at Naval Heritage Ship Museum. Portsmouth, UK visit 1993
(Source: Author's own records)**

UN Peacekeeping Assignment
in Bosnia Herzegovina

I had barely spent six months at Rangpur Brigade as
Officiating Brigade Commander when I received a call
from Chief of General Staff at Army Headquarters

telling me that Bangladesh Army has committed to send a mechanized infantry battalion to Bosnia Herzegovina, and that I was being assigned as the contingent commander. Having done a reconnaissance of the likely deployment area in Bosnia the previous year, I felt the army may not have given serious considerations to the challenges to deploy a mechanized battalion in the given tight time schedule. To us, as non-Europeans, there were serious issues like the Yugoslav wet and cold winter, and the fact that Bangladesh Army did not have winter clothing like thermals, appropriate winter jackets for freezing conditions, and most importantly did not have winter boots for the soldiers.

Bangladesh Army is primarily equipped for warm temperate equatorial regions, and Bangladeshi soldiers had never seen anything like snow. I still wonder how Bangladesh A rmy could decide to deploy a mechanized battalion which in reality did not exist in the Bangladesh Army in 1994. I believe the policy being not to turn down any request from UN to deploy Peacekeepers and possibly that's how the decision was taken. I had earlier done the ground reconnaissance in Bosnia in 1993 and knew that we were expected to replace a highly capable and experienced French Mechanized Battalion. So, the UN expectations from the incoming Bangladesh battalion would be high, they would expect the Bangladesh Battalion to perform like any other NATO unit and we were nowhere near that capability except for our soldiers' resilience and determination to carry out the task.

The designated contingent was to be organized around 10 East Bengal, *a non-mechanized infantry battalion*, as the core unit. Supporting elements that would be added for logistical support were the medical detachment including a surgical and dental team, communications detachment to provide us radio and other communications support, a workshop unit to take care of all the vehicles specially the armored personal carriers—APC and supply unit to stock and cater for our food and fuel needs. My experience as Peacekeeper all through the Bosnian War had been spread over fourteen months across the conflict period 1994 -95. Further the UN Peacekeeping mission in Bosnia was in the midst of many controversies with the media reporting only the information they had access to. Many events went unreported to the outside world because the UN press releases were tailor made to suit the agenda of the decision makers. Therefore, I am separating the Peacekeeping story for a later part of the book as a complete chapter where I will discuss the issues in detail.

I returned from UN mission in November 1995, and after maybe two weeks of debriefing at Army Headquarters, I received my long due promotion to the rank of Brigadier General. Interestingly in my military career all my promotions came delayed. While professionally I had always done well on my military courses, always achieving positions among the top few if not the top position, had good career placements, served as instructor in all ranks from captain upwards, but yet when it came to promotions there always seemed to be something holding me back. On my career path I had two bumps, the memory of which still makes me feel bad. First was in 1984, after I finished my yearlong staff course at Mirpur, and I topped my course, that would mean based on results I finished at number one position. At staff college the practice is not to officially tell the students about the positions in any formal way, however. A few years later when I was working as the secretary to the army chief, I reconfirmed my staff college results from the military secretary branch where the personal dossiers and records are kept as confidential documents. At the time it was customary practice in Bangladesh Army that the officer who tops the staff college course would be sent for second staff college generally to USA. For whatever reasons I was not sent for the second staff course abroad. The second professionally upsetting incident happened in 1998, when I was selected for attending the National Defense Course (NDC) in India and my name had already been forwarded to NDC India and much of the official formalities were complete. After handing over the command of my Bandarban Brigade, I was attached to Army HQ Dhaka to complete the medical requirement for overseas travel. At the last moment, my name was dropped and another officer had himself nominated. After cancellation of my overseas course selection, I was re-directed to go as Brigade Commander to Bogra Brigade instead of a course abroad. Both the incidents left painful scars, because on both occasions my professional ability got me to the deserved position, but perhaps I did not nurture contacts, which I guess became the overriding factor.

Commandant Bangladesh Military Academy

After my return from the UN Peace Keeping Mission in Bosnia, I joined the military academy as commandant in November 1995.

Author on left, Commandant. Bangladesh Military Academy, Chittagong, June 1996 (Source: Author's own records)

To be posted as Commandant of the Military Academy is considered as a prized posting and recognition of one's professional ability. So, I was indeed very happy on getting this rare opportunity to be the commandant of the military academy. A military academy generally runs on a very set routine and major events go as per calendar and standard operating procedures (SOPs), which have evolved over years. It was impressive to see the events run like a clock and people at different levels guided by SOPs. My concern was therefore on non-routine activity.

On the very first year I noted that a number of cadets had been demoted to the lower batch or junior batch and in some cases had been withdrawn from the academy due to injuries incurred while training at BMA. I found this a bit disturbing that we were ruining a young man's career, probably unnecessarily. Most cadets came from a background where they never participated in outdoor activities or sports to develop their physical fitness like running. Cadets who came from cadet colleges had no difficulty in facing the physical rigors because they had trained into high standards. I explained to the instructors that if our training system caused injuries making the trainees unfit, than we had to review the system. My simple logic that I presented to the instructors was this: *"if I give you 100 cadets to train, you have to return back to me 100 cadets after training, system loss indicates to me system inefficiency and needs review of the practices."* I think it was from 37th BMA Long Course that I implemented the practice of slowly building the pressure instead of applying "shock and awe" on Day 1. Physical activity pressure was gradually scaled up to allow the weak ones to build their muscles and stamina. With the new practice, cadets started doing run-walk-run (*RWR*) in the first two weeks in soft running shoes, not military boots like before. While my experiments did upset some of the military instructors or platoon commanders, as we called them, in the end we did completely eradicate the injury related relegations, probably saving the careers of many individuals.

The year after, I brought major changes to attitude and basic practices in weapons training. Generally, perception in the military about weapons training is all about ragging. So, in weapon training classes the attitude of instructors was with mind set to shout and award physical punishments. From my own experience I knew cadets dread weapons training classes and would be happy whenever these were cancelled. My suggested change was based on the concept for instructors to spell out clearly

'Class Objective for the Day.' I told them if the objective for the day was to build physical fitness, then go ahead with punishments or ragging. But if the class objective for the day was to teach weapon handling or to improve firing skill, then create conditions to remove that fear of weapons training. As a junior officer I was a weapons instructor at the army school, so I was very confident about anything I said on weapons training. When I compared my personal experience on weapon training, I can say that I learnt very little at the military academy due to ragging and punishments. I learnt weapons very well at Infantry School in India sitting on a bench under the shade of a big banyan tree, without the fear of punishments. The changes that I introduced for weapons training did produce marked improvement in the firing skills.

My observation of more serious concern, was that the cadets were being groomed to perform under supervision only, *'zero mistake syndrome.'* In times of war military officers have to learn to take risks and accomplish the mission under 'fog of war' and without direct supervision, just keeping the end objective in mind. Especially under electronic warfare environment, command communication network may totally be disrupted. Such a situation will demand leadership, initiative and quick decision making and that is what makes you win or lose the battle. While the practice of *'wait for orders'* maybe essential to play safe to avoid failure and may be a good practice for non-military management, but it is certainly not always ideal for military leaders.

I was uncomfortable with the *'zero mistake syndrome'* now deeply imbedded in most practices at the military academy. I understand in peace time army, one gets punished for mistakes and this may also affect one's career path, so everybody plays safe. The '*zero mistake syndrome*' led to over supervision of the cadets by their platoon commander or instructors on almost every step of their activity. Even during evening self-studies in dormitories military instructors were visiting to ensure that cadets are doing their room self-studies. To me we were treating the cadets like minors and not adults. I often told the instructor platoon commanders that we had to learn to treat the cadets as adults. I feel that over the years many of the cadet college culture had crept into academy practices.

My thinking was that the final semester cadets will transition into commissioned officers in a few months to take on their responsibilities, so they need to be given opportunities in the military academy to practice their future role. Therefore, as commandant, I insisted that final term cadets be given opportunities to practice leadership and supervision responsibilities. I introduced the practice of all dormitory related supervision of junior cadets to final term cadet appointments like the cadet platoon sergeant or corporals. This in my view, gave an opportunity to final term cadets to practice taking responsibilities. The toughest challenge for me was to delegate the responsibility of umpiring and refereeing for competitions also to final term cadets. I motivated the instructor platoon commanders by telling them that in few months' time these cadets as newly commissioned young officers would be doing similar jobs in units so why not let them practice even if they make mistakes. We settled on the decision that boxing and assault course being of more serious nature will continue to be supervised by military officer instructors. All other competitions like debating was to be supervised by designated final term cadets, a big departure from past years. Getting away from '*zero mistake syndrome*' mentality is indeed very tough for most organizations. Over a period of time happily my proposed changes were implemented, but I am not sure whether the practices were later reversed or continued. My concept was final term cadets need to be given more and more responsibilities which allows them to develop their leadership and decision making qualities.

I witnessed the results of '*zero mistake syndrome*' mentality about two years later when I was in Bogra as a Brigade Commander. Army Chief, General Mustafiz, came to visit my brigade during a field exercise. The Army Chief asked a junior officer what he will do when he get fired from enemy, the junior officer quickly replied I will inform my company commander and wait for his instructions. The Army Chief looked at me and I am not sure what he wanted to say but kept calm. I had to quietly tell him, 'Well that's how the military academy is training the cadets' to wait for orders. To me this is certainly not the way to go about on a battlefield. Junior commanders have to learn to take tactical decisions at their level keeping the end-objective in mind and also know their limits when he needs to look back for orders to go further.

Bandarban Brigade

In March 1997 I was posted from the military academy to Commander of Bandarban Brigade in Chittagong Hill Tracts. The brigade was involved in counter insurgency operations in south, mostly along the Myanmar border. My stay at Bandarban and the knowledge gained on insurgency in the area was very helpful years later when I was serving as Bangladesh Defense Attaché in Myanmar. I had the unique opportunity of travelling along the border on both sides.

One of the important incidents of my time that I can remember was a daring raid by insurgents on border security force outpost at Thanchi Bazar. The outpost was located on the riverbank on the southern edge of the bazar. The main camp was some two hundred yards away on the hilltop. This outpost was held by around ten soldiers with the purpose to prevent insurgents from disrupting the bazar activities and more importantly to protect the fairly large number of civilians staying in the bazar area including their shops which provided essentials to the locals. In 1997 the only way to reach Thanchi was to travel by river boat, but I understand now we have all-weather road going even beyond Thanchi. The security outpost on the riverbank was raided during early hours may be around 4am which resulted in death of one of the soldiers. There was only one track from the river side to enter the bazar which this outpost was guarding. As a standard security measure for the night, one LMG (*light machine gun*) use to be placed for fixed line fire covering the track to the river with two soldiers on duty manning the machine gun. It transpired later that the insurgents had planned for a silent raid to grab and kill the soldiers at the outpost. The insurgents probably had come to know that most of the on-duty soldiers remain at sleep during the night, so this was good piece of intelligence for planning the raid.

We were saved by sheer luck of an incident occurring in violation of the standard procedures. One of the on-duty soldiers left his duty post on that night of the incident and went out towards the river for early morning nature's call, but the good part is he cautioned the other soldier on duty with the LMG to remain alert and watch the front to the river. This was about the time when one of the insurgents had already reached a position just about twenty yards from the duty post undetected. The soldier who had gone out bumped into this insurgent and they both grabbed each other both shouting and trying to overpower the opponent. The soldier on duty at the outpost simply squeezed the trigger of the LMG which was facing in that direction along the track.

Long bursts of automatic fire from the LMG killed both the insurgent and the possibly the same fire killed our soldier also. The other soldiers jumped out of their beds took up positions and started firing in the direction of the suspected insurgents. Exchange of fire continued for some time as the rest of the insurgents withdrew into the night darkness. In the morning both the dead bodies were found next to each other with multiple bullet wounds. It was indeed a very daring raid being conducted by the insurgents. It was simply good luck that one of the soldiers had gone outside the post and bumped into the insurgent which saved the lives of other ten sleeping soldiers. Quick response from the other young soldier on duty who fired the LMG non-stop saved the situation. Had the insurgents been successful to dispose the two LMG crews, other soldiers who were sleeping at the time would have been all been killed that night. To me this was classic example of 'silent sentry disposal' which was taught to us at the military academy.

As a Brigade Commander I had a helicopter at my disposal next morning so I flew into the area to investigate and also sent out a number of patrols to hunt for the insurgents who were not to be found. Later I had sent a citation for the two border security force soldiers for award. I am not sure whether these awards were approved or not as the process takes a long time. There was another incident of great interest to me that I can still remember in great detail. This was about raiding a large insurgent training base in the hills east of Naikhongchhari Upazilla. A local intel officer came to brief me and told me that his sources were telling him about the presence of a very large insurgent camp about ten kilometers in the hill forests east of Naikhongchhari Upazila. Hearing the description, I told the intel officer that if his report was correct, we should be able to easily see the camp from helicopter to which the intel officer agreed. So, we decided to do a helicopter reconnaissance of the forest area and we combined this trip with the monthly ration drop helicopter sortie scheduled for the month. The helicopter sound usually gives away the plans. So as part of the tactical routing we planned to fly south of Thanchi along River Sangu to visit the Mowdak border camp on the southeastern border close to Myanmar border. After visiting the border guards camp at Mowdak, we flew west over the Chimbuk hill range and across Alikadam valley straight towards Naikhongchhari.

Soon after crossing the Chimbuk hill range on its western slope, we spotted a number of huts with blue plastic sheets. This for sure was not a tribal village, because tribal people do not use plastic sheet for huts. Clearly the settlement was non-tribal and use of plastic sheets were not natural to the area. It was a fairly large settlement like a hole carved into the hill forest. The ground inside was cleaned and shaved of bushes and trees and therefore brown clay surface was visible from the air. Army patrols walking in the surrounding area at ground level would not see this because the surrounding deep forest screened the location. We immediately spotted some people in dark clothes running around with weapons. The pilot pulled up and away from the area to avoid hostile fire. Standard procedure was for the helicopter to remain above three thousand feet as a safety measure against small arms fire. Later I organized a large force to raid the suspected area only to find the camp site already abandoned. I guess the insurgents left the site after our helicopter recce flight.

While I was in Bandarban Brigade I was selected for the NDC course in India. I handed over the brigade command responsibility to the new incoming brigade commander. Thereafter, I was attached with Army HQ in Dhaka to complete overseas travel formalities. May be about fifteen days before my flight my name was dropped and I was posted as brigade commander in Bogra. I had travelled from Bandarban to Dhaka, but after about a month stay instead of proceeding on NDC course abroad I travelled with my family to Bogra. I stayed with the brigade in Bogra for one year as I was nominated for the Bangladesh military first NDC course at Mirpur Dhaka. This being the first course the institution was not fully settled, the first batch students coming from outside Dhaka were neither provided accommodation nor transport except for pick and drop for class schedules. I took this one year as a time to focus more on building my house as I was getting close to my retirement. After finishing the yearlong NDC course me and my family were excited to learn that I was posted as Defense Attaché to Myanmar. This placement abroad came at about a time when officers of my seniority would come up for promotion consideration. On completion of my tenure as Defense Attaché I would be returning with a year left into my retirement. Therefore, the placement for three years away from the mainstream army was also an indication for me that I would not be in consideration for the next promotion. While some people were happy to see me placed abroad for three years making it easier for some of my juniors to get promoted. Honestly speaking I understood the political realities very well and had mentally accepted to start planning for post-retirement life. Therefore both me and my family were happy to have been posted abroad at the time as it helped me financially to support my eldest son, Ehtesham, who was studying at the time in University of Arizona, Tucson, USA. Ehtesham is now a US Citizen and works for Amazon as a Data Scientist.

Myanmar 3 Years as Defense Attaché

I worked as Defense Attaché in Myanmar for three years, from 2000—2003. Serving as a Defense Attaché is an excellent opportunity for military officers to experience a life centered around diplomatic circles, a world away from the military's monotonous and disciplined lifestyle. Attending receptions, dinners and golfing were all part of the job. Making trustworthy friends both among the diplomatic community and the locals is what makes a successful tenure for a Defense Attaché. Personally, my family and I greatly enjoyed our three years of stay in Myanmar. On the social side, we made many new friends from different countries, and some still keep in touch.

Author at far left, meeting Myanmar Army Chief, General Maung Aye, 2000

We made quite a few Burmese friends, we found the Burmese people very nice, good and friendly community. During my stay in Myanmar, I particularly enjoyed golfing, which is very popular in Myanmar, especially with the Myanmar military. In every township you visit, you are likely to find a golf course because golf is a very popular game, and land is abundant in Myanmar with a comparatively small population. Since its independence in 1948, the Burmese military has been fighting insurgency all over the country. Wherever the military camped, they built a somewhat playable golf course which developed over time. Since 2000, some upscale golf courses have also been built with private investment. Golf is the only surviving trace of past British colonial legacy apart from the colonial time grandeur buildings in downtown Yangon. One of the oldest golf courses in the region is the 'Rangoon Golf Course', which is more than 120 years old now. The course has been renamed and is now called the 'Yangon Golf Course'.

Author with Aung San Suu Kui at US Embassy Reception in Yangon, Sept. 2002

Interestingly, while the Myanmar military is in love with golf, it hates the game of cricket, which was possibly introduced during British colonial times. The game of cricket was never picked up by the local Burmese people. Cricket is now played in Yangon only by a small Indian community that probably represents the third generation of immigrant settlers. I was told the Myanmar government had once even officially banned cricket playing.

The term Indian community in Myanmar would include people whose forefathers came from India and Bangladesh. Visibly the settlers and the Burmese are clearly two different races. The settlers are of a darker shade with sharp features like the Indians, while the

Burmese generally are of brown skin and visibly of
Mongoloid race. Indian communities are mostly found
in downtown Yangon and in the neighboring townships.
Smaller groups of Indian settlements can also be found
even in some remote townships close to the Indian and
China border in the north.

Author in foreground visiting Myanmar Insurgent Administered
Area close to Thai border. Insurgent soldier holding umbrella for
the author with Myanmar Army Officers seen in the distance, 2002
(Source: Author's own records)

In the 1940s and 1950s, there used to be regular sea route steamer service between Chittagong and Rangoon. Many business entrepreneurs from Chittagong, Bangladesh, had settled in Yangon during the British colonial period for business or jobs. Many of these people married local Burmese women and had a second family. My maternal grandfather (*nana*) lived in downtown Yangon. He worked with the Chittagong—Rangoon steamer service and also had a small business in Rangoon. From the old family records, I did manage to identify his house address in downtown Yangon which appeared to be occupied by some banana merchant now. He left everything and fled back to Chittagong in 1942 as the Japanese approached and captured Burma. This grand old man, my maternal grandfather (*nana*), had his second and third wives from Burma and later brought these ladies to Chittagong. I met both of my Burmese step-grandmothers (*nani*) as a small boy when I used to visit our village home during school vacations in Hathazari Chittagong. One was from Mandalay, and the other one was from Yangon. Both the ladies died of old age and are buried in Chittagong.

During my stay in Myanmar, I made another interesting discovery, many of the renowned businesspeople from Chittagong have old family ties in Yangon. Their elders had married in Burma in the 1940s, so now many Chittagong families have blood cousins living in Myanmar, particularly in Yangon city. The most interesting story is of Ambassador Rahim, the Bangladesh Ambassador to Myanmar, in 2002—2004 during my time. His father lived in Yangon and had married a local Burmese lady. I am told that Ambassador Rahim was born in Yangon to his Burmese mother. His father later returned to Chittagong in 1960, along with only one son, Rahim, while the rest of the family continued to stay back in Yangon and still live there as Myanmar citizens. So, the young boy grew up in Bangladesh and later joined the foreign service in Bangladesh. He speaks fluent Burmese and was with Bangladesh High Commission in Yangon twice, first as a Counsellor and later as Bangladesh Ambassador to Myanmar in 2002. An incredible story for a boy who was born in Yangon to a Burmese mother to have come back to Myanmar years later as Bangladesh Ambassador. I have met many other Bangladeshis from Chittagong who have such family ties in Myanmar.

The job of Defense Attaché is to keep an eye on the military developments in the host country. Within the Defense Attachés community, we had subgroups with common interests. Defense Attachés would meet in small groups to share information and discuss Myanmar military activities. However, sharing vital information depended absolutely on person-to-person friendships. Diplomats and the Defense Attaché's community preferred being on the golf course for informal discussions and meetings. In a country in complete control of the military, such as Myanmar, one expects most hotels or restaurants frequented by diplomats to be bugged or the local staff on listening watch. The golf Course was the safest place, too big a property to be bugged.

In Myanmar, foreign visitors and tourists are allowed to visit only a few cities outside Yangon, like Mandalay, Bagan and Inle Lake in central Myanmar. Most of Myanmar is closed to travel by foreigners. Visitors require special permission from military intelligence to visit areas other than the few specified tourist cities. Every year Myanmar military intelligence organized trips to various parts of the country for the defense attaché and their families. These visits were an opportunity for us to visit different military establishments and some remote townships, like our visit to Muse on the China border. My visit to Bagan and staying overnight at a hotel on the banks of the Irrawaddy River is unforgettable experience. Bagan, with hundreds of pagodas in decay now, was built a long time in the past. It was the capital of the old Burmese Kingdom and a place one must visit. Being from Bangladesh, my job and my major area of interest was to keep track of developments of Myanmar Military, its weapons acquisition programs and specifically military activities in the northern Rakhine State bordering Bangladesh. Myanmar has a small border with Bangladesh, a combination of a strip of land and the NAF river boundary.

Bangladesh has no border disputes with Myanmar, and the borders are clearly demarcated. I was involved with the border demarcation pillar setup, and the border demarcation was completed in 1982. The only dispute with Myanmar is about the Rohingya people who mostly live in the Rakhine State townships bordering Bangladesh. Other than the bordering townships rest of the Rakhine state is populated by Rakhine people of mongoloid race. Interestingly, Rakhine people are also found to be living on the Bangladesh side of the border up to Cox Bazar and even have settlements in Pathuakhali, Barisal. Many of the villages south of Cox Bazar still carry Burmese-like names, possibly because a few centuries years back Rakhine Kingdom was an independent Kingdom which possibly included the southern areas of Cox Bazar.

My three years in Myanmar, travelling and meeting people across the country helped me understand this unique and interesting country. Myanmar is heterogeneous and consists of diverse ethnic groups speaking different languages who live in the hilly fringe states surrounding the central plains. These ethnic groups, such as the Karens, Chins and Kachins, and many other smaller ethnic groups, see themselves as different from the Burmese people or the Burman's, the people from the central plains. These ethnic groups have long resisted central rule from Yangon and did not want to be part of the newly independent country of Burma. Burma, later renamed Myanmar, gained independence from the British in January 1948. Soon, the country plunged into a nationhood crisis, with the different ethnic groups refusing to become part of the Union of Burma. So, from the day it gained independence, Myanmar faced the challenge of keeping the Union of Burma together as one country. Myanmar military had to involve in military operations to fight different armed ethnic groups all over the country. Ethnic groups living along the eastern borders of China, Lao and Thailand inhabited areas where the drug trade flourished, generating millions of dollars in revenue. This drug revenue helped these ethnic groups to arm and maintain a fairly large army to fight the Burmese military.

Along with separatist movements in other parts of the country, on the Bangladesh border, Myanmar had to deal with the Rohingya population showing dissent to the Myanmar government. Myanmar's government considers Rohingya people outsiders who took advantage of the British colonial administration and settled in the region. However, the Rohingya history completely differs from the Burmese view claiming that the Rohingya people lived in the area long before the British colonized Burma. In the 17th and 18th centuries, nation-states did not exist with firmly defined borders as we see them today, so the notion of a nation-state of today would not hold good for the political maps of 17th and 18th centuries when small independent kingdoms governed the area comprising of today's Rakhine or Arakan State. Going back into Burmese history, we find an interesting fact that Arakan, later renamed Rakhine, was always an independent kingdom captured by the Burmese King sometime in 1784, so the Burmese themselves are the alien masters of the Rakhine or Arakan State. Burmese rule over Rakhine lasted about a hundred years until 1826. With the defeat of the Burmese in the First Anglo-Burmese War, Rakhine was ceded to Britain, and thus Rakhine or Arakan State came under the British Colonial administration.

Rohingya's living along the north-western Rakhine state bordering Bangladesh started their own armed struggle like in other Myanmar states. Since the Rohingya population inhabited mostly the four townships in the north bordering Bangladesh therefore, the Rohingya insurgency was localized to these townships only. The major part of the Rakhine state, the center and the southern part, were not affected by Rohingya insurgency simply because these areas are inhabited by other ethnic groups, the Rakhine and the Burmans. While Myanmar Army countered the armed uprisings the Myanmar government continued their efforts to negotiate for peace to accommodate the demands of the different ethnic groups. Myanmar peace negotiations with Rohingya leaders came close to settlement in 1991, a time when most other insurgent groups had already signed a peace agreement with the Myanmar government. However, the peace agreement effort between the Myanmar government and the Rohingya failed.

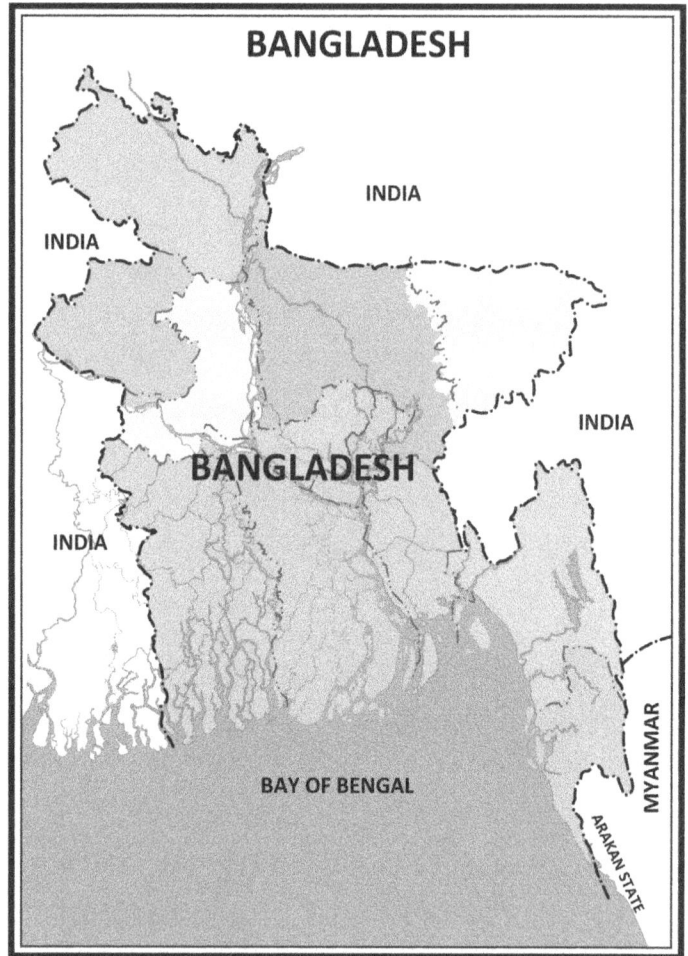

Copyright: GEO CONSULTS, Bangladesh

After its independence in 1948, three elections were held in 1951, 1956 and 1960 for both houses of the Myanmar Parliament, and five Rohingya muslim leaders became 'Members of Parliament' with one even becoming a minister. This clearly shows that Rohingya people were very much considered citizens of Myanmar. After the 1962 military coup, Myanmar came under direct military rule, and it may not be wrong to say that General Ne-Win, the coup leader, created the current Rohingya crisis by stripping the Rohingya people of citizenship, thus making them stateless people within their own country. Today, Bangladesh hosts some of the world's largest refugee camps for Rohingya people who were driven away from their homes by the Myanmar military. In pursuance of my interest in Northern Rakhine State and making use of my golf friendship, I did get permission once to visit the old city of Akyab, the capital city of Rakhine now known as Sittwe. Sittwe became famous during World War II as the newly built airport became an important staging base for British Army and the Royal Air Force.

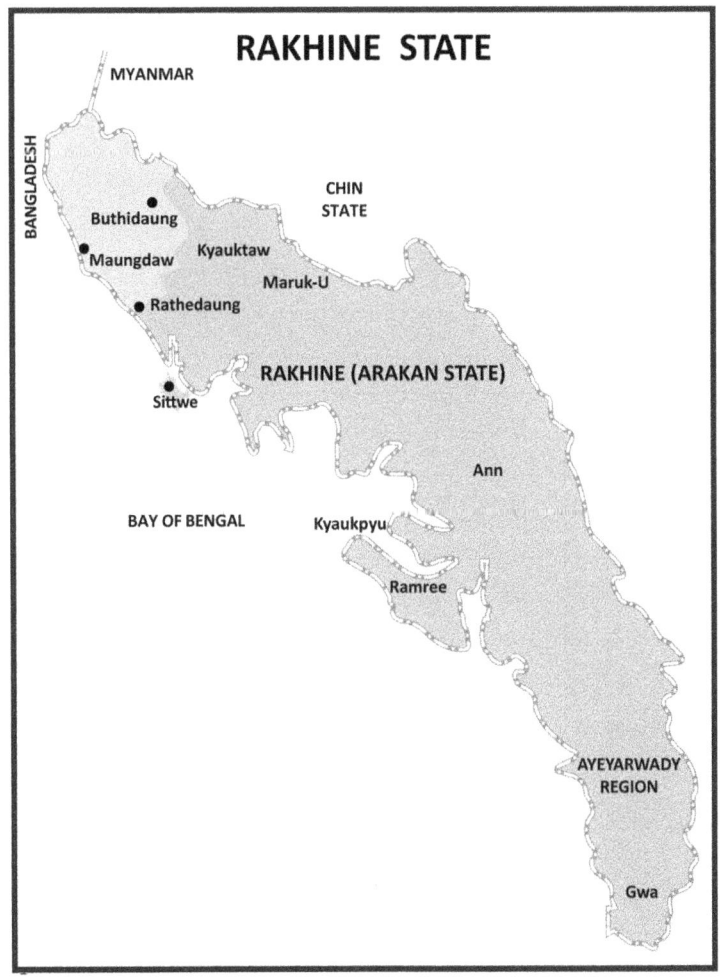

Copyright: GEO CONSULTS, Bangladesh

Bangladesh has a counsellor office in Sittwe with a small staff. For my visit, I flew into Sittwe on a commercial flight on which I was the only foreign passenger because Sittwe is not open for visits to foreigners. Local flights to most cities in Myanmar serve locals only. Locals pay a small, subsidized fare in local currency, while I, as a foreigner, paid for the ticket as regular fares in USD. Luckily, Counsel General Emdad at Sittwe was a known old colleague, so he took good care to make my stay and trip most comfortable and rewarding. I stayed overnight at Sittwe with Emdad, where the city was provided electricity only for a few hours in the evening with generators. The next morning, I boarded a passenger ferry boat for a day-long journey to Buthidaung. This passenger ferry is like the one in Bangladesh that runs between Dhaka—Barisal. The crowded ferry left Sittwe river port at about 10 am and reached the Buthidaung around 4 pm. I travelled as a guest of Myanmar military intelligence, who provided me with a jeep and escort.

A foreigner would not be allowed to travel in this area unless cleared and escorted by military intelligence. After about an hour's drive, we reached the Myanmar side of the bank of the Naf river just before sunset. I could see the lights on Bangladesh's side across the river in the far distance. Here the mid-stream on the Naf River denotes the international boundary between the two countries. For me, as a student of military history, it was an exciting experience to be driving on the Maungdaw road towards Naf River, which had seen some fierce battles during World War 2 between the Japanese and British forces. I did see some of the remains of World War 2 bunkers of the Japanese Army. Some of the concrete machine gun bunkers were still in reasonably good condition along the roadside. We were not allowed to stop, but it was enough to have the opportunity to drive through the battlegrounds and the old Japanese Army battle positions.

Brig. Gen. Salim Akhtar (author) and Col. Dibrel, USA being greeted at Kachin State Airport, north of Myanmar

Brig. Gen. Salim Akhtar (author)—front row, far left—in Myanmar National Day Parade in Yangon

The next day I had to retake the day-long river route journey back from Maungdaw to Sittwe. From Sittwe, my return journey to Yangon was exciting as I had decided to take the coastal high-speed passenger boat along the coast into central Myanmar past the deep seaport location of Kyaukpyu. We landed at a river immigration point on the west coast somewhere in central Myanmar. I was the only foreigner among the passengers; the immigration police checked my passport and allowed me to proceed. We had a car on rent which took us to Yangon, another four hours drive. My decision to travel by coastal river boat was risky, skirting the shores along the Bay of Bengal. I travelled on the coastal boats to have a close view of the much-talked-about deep-sea port at Kyaukpyu port which China was building. Kyaukpyu port (pronounced as 'chowk-piu') was on global news at the time, and I was happy to be able to provide a first-hand report on the status. Kyaukpyu deep seaport is functional now, helping China to bypass the Malacca Straits in its shipments. Now China can also transport its imported oil from the Middle East through the oil pipeline, which connects Kyaukpyu deep sea port to China. I returned from Myanmar sometime in Apr 2003, and with a year of service left, I was posted as a Director at Army Headquarters. I retired a year later, on 30 March 2004, at the age of 53 years, which was a very young age to retire.

End of Military Service

I retired at a young age with the vast experience I had acquired, and I certainly could have given some more useful years to the Army. However, after retirement, I took a break from the routine work schedule that I had followed for nearly thirty-five years. For one year after my retirement, I did nothing except playing golf twice a day. In 2007 I got a wonderful opportunity to do a voluntary job for the golf federation. For the next ten years working for golf became my passion. I spent long hours connecting to USGA (US Golf Association) and R&A (Royal & Ancient) in Scotland. I got USGA to send a team to do the first-ever course ratings for golf courses in Bangladesh. The other area I focused on and worked on was developing a close working relationship with R&A (Royal & Ancient), a global golf governing body based in St Andrew's, Scotland, UK. I am grateful to R&A for funding projects I initiated to develop the game of golf in Bangladesh. My focus had been to select boys and girls from humble backgrounds and pay for their education as they learned to play golf. With a great deal of happiness, I now see many players who play on the Bangladesh golf professional circuit as products of my golf development programs, known as the R & A Training Program for the underprivileged. This golf program helped these poor boys and girls make golf a career besides completing high school. I feel so happy about providing educational support to so many girls and boys in my work with golf.

However, sitting at home and doing little soon became boring. I was in good health and could still work rather than sit at home. So, in 2005, I grabbed the opportunity to join a private company called G4S. This second profession for me lasted for twelve years. G4S is a multi-national security services company with its Head Office in London and has a presence in more than a hundred countries. In G4S, I was Head of Manned Services and was later assigned the additional responsibility of looking after the cash-carrying and Electronic Security services business. I retired from G4S in 2016, reaching the retirement age of sixty-five, a corporate retirement age. I was always very passionate about working with computers and IT, so in 2017 I established my own software company called IBSL (Intelligent Business Solutions Limited—www.ibsl-bd. com). The software company still exists, but I have slowed down and decided to spend family time with my sons in USA and Canada. While I spent my life working, my wife spent all her time looking after and grooming our three wonderful sons. Private schooling was expensive, but my wife was very firm in her objective to put all the kids through good private schools. Whenever my workplace was away from Dhaka, my wife and mother took care of the kids, ensuring their schooling would not be disrupted. We sacrificed little of what we possessed but ensured good schooling in Bangladesh and later university study in the USA. We are proud as parents to see that all three sons are well-educated and are happily settled as engineers & scientist in USA and Canada. As I look back on my life, I can count on three distinct professional paths I pursued during my lifetime. I started with a military career at the tender age of nineteen years, and this path lasted for about thirty-five years. After retiring from the military in 2004, I stepped into my second career in 2005 with the private security services company called G4S. Finally, in 2017 I stepped into my third career, establishing a company pursuing software-related business. At the age of seventy plus as I write this book, I spend my time playing golf and working as a consultant for golf projects.

UN Peace Keeping in Bihac
Bosnia Herzegovina

Legend for Chapter 7
War in Bosnia Herzegovina

1. **Krajina Serbs / Croatian Serbs Army—RSK**

   RSK (Republic of Serb Krajina) *a* self-declared republic in control of areas in southern Croatia bordering Bosnia Herzegovina during the war.

   RSK operated as allies for Bihac Rebel Muslim leader Fikret Abdic and Bosnian Serbs. RSK military units had encircled Bihac enclave from the north and the west.

2. **Bosnian Serbs Army—BSA**

   BSA was armed and supported by the Serb Army, former JNA units and Serb militias from Serbia. BSA encircled Bihac from the south and east, thus in coordination with RSK had completely cut off Bihac from all sides. General Ratko Miladic was the commander of BSA. After the war he was tried and convicted for war crimes.

3. **Serbia Army—JNA**

   Serb military units of former Yugoslav Army which became part Republic of Serbia after secession. They inherited all the heavy weapons of the former Yugoslav Army and actively supported both the BSA and RSK.

4. **Croatia Army—HV**
   Croatian Army formed after Croatia declared independence
   from former Yugoslavia and became an independent country.

5. **Bosnian Croat Army—HVO**
   HVO was formed by the Croats in Bosnia Herzegovina
   and had active support from Croatia.

6. **Bosnian Army—BiH**
   Bosnia government military units and represented
   primarily the Bosnian Muslim forces.

7. **Bosnian Government military forces in—5 Corps**
   Bihac which defended Bihac against Serb attacks
   throughout the war. General Atif
   Dudakovic was the commander of 5 Corps.
   He later became the Army Chief in independent Bosnia.

8. **United Nations Peace Force Headquarters—UNPF**
   UNPF with HQ in Zagreb, Croatia controlled UN
   peacekeeping forces in Croatia, Bosnia, Herzegovina
   and Macedonia.

9. **United Nations Protection Force with HQ in Sarajevo
   central Bosnia—UNPROFOR**

10. **Bihac Areas Command—BAC/BH Comd**
    A new setup created in Bihac after arrival of Bangladesh
    Battalion.

11. **Combat Air Patrol—CAP**
    NATO aircrafts flew the CAP mission in Bosnia
    Herzegovina during the war and were available for air
    support mission when approved by UNPROFOR HQ.

12. **Bangladesh Battalion—BANBAT**

13. **French Battalion—FREEBAT**

14. **Armored Personnel Carrier—APC**
    Vehicle used for carrying 8 -10 soldiers into combat
    providing protection from small arms fire. BTR 70 was

the Russian APC model used by Bangladesh Battalion in Bosnia Herzegovina.

15. **United Nations Military Observers—UNMO**
Unarmed UN military officers placed in the conflict zone to report ceasefire violations.

16. **Religious Divide in Bosnia Herzegovina**
Bosnian Serbs—Orthodox Christians,
Bosnian Croats—Catholic Christians,
Bosnian Muslims -Islam

17. **Fikret Abdic**
Muslim rebel leader from Bihac who had declared the North Western part of Bihac enclave as an independent and autonomous region. Allied with Serbs and fought the Bosnian Government forces but became a refugee after his forces were defeated.

18. **Daily Situation Report sent in specified format— SITREP**

19. **United Nations Security Council Resolution—UN SCR**
Document that incorporates the decisions of the UN Security Council.

20. **UN Protected Area—SAFE AREA**
Provides physical security to unarmed civilians within the specified area. When established UN Forces are structured to provide the deterrence and protection.

UN Peace Keeping in Bihac Bosnia Herzegovina

Seeds of Conflict

The brutal war that raged in Bosnia Herzegovina, the associated ethnic cleansing, the genocide and the disintegration of former Yugoslavia should not be viewed as an event that occurred at the spur of the moment. In former Yugoslavia, the seeds of the conflict between the three ethnic communities, the Serbs, the Croats and the Bosnian Muslims, are deeply rooted in Balkan history. Events of the 1980s and 1990s in Europe, the death of President Tito of Yugoslavia, the breakup of the former Soviet Union, the falling of communist regimes in Eastern Europe, and finally, the economic downturn in Yugoslavia in the 1980s are all contributing factors in rekindling of smoldering fire of the inter-ethnic hate.

Austro-Hungarian Empire and later the Ottoman Empire ruled this region for most of the 14th century up to the early 20th century. Both the Austro-Hungarian Empire and the Ottomans during their rule favored one community over the others, thus subjecting the non-favored community to a disadvantage, resulting in inter-community hatred. Serbs carried grievances against the Ottomans for alleged mistreatment, which turned into anti-muslim sentiments over the period. Serbs also blame the Ottomans for converting their cathedrals and churches into mosques across the Balkans. Slobodan Milošević, the then President of Serbia, even publicly stated this as the Yugoslav conflict unfolded. The motive for making such public statements by Slobodan Milošević, President of

Serbia, can be better seen by the outcome of the inter-community fighting and carnage that followed. The allegations may or may not be true, but they helped fuel the anger and hatred against the Bosnian Muslims in the 1990s as the inter-ethnic fighting unfolded.

During the long rule of the Ottoman Empire, the population of the Muslim faith grew significantly, particularly in the region now identified as Bosnia Herzegovina. Ottoman's offered significant benefits for conversion to the Islamic faith, which may have encouraged people on the survival threshold to convert to Islam to avoid persecution and to gain favors from the ruling class. Yugoslavia was created in 1918 at the end of World War 1 when the victorious Allied Powers split up the territories of the former Ottoman Empire to create new states. Serbs had long dreamed and demanded a country for the Southern Slavs in the Balkans. The term South Slavs generally refers to Slav people inhabiting the Balkans. During Second World War, allied forces, particularly the UK, worked closely with the Yugoslav partisans, which helped defeat the Germans in the Balkans. Therefore, the Serb demand for a country of their own had support among the Allied Powers, particularly with the UK. Yugoslavia was thus created in 1918 and was named the 'Kingdom of Serbs, Croats and Slovenes'. It's interesting to note that there was no mention of the presence of the Bosnian Muslim population in the declaration, and that must have undermined the confidence of the Bosnian Muslims in the new state of Yugoslavia.

The region that encompassed the state of former Yugoslavia held several independent Kingdoms earlier, which reflects the ethnic diversity in the region. The earlier dynasties in former Yugoslavia territories were the Kingdom of Serbia, the Kingdom of Montenegro, the

Kingdom of Croatia and the Kingdom of Bosnia. The new state of Yugoslavia created in 1918 was centered more around the 'Kingdom of Serbia' with it's seat of power in Belgrade; therefore, the government was inherently Serb-centric. Peter, the King of the Kingdom of Serbia, became the first king of Yugoslavia until he died in 1921. Peter was succeeded by his son Alexander as the new King. King Alexander renamed the country the 'Kingdom of Yugoslavia' in 1929 and decreed that all citizens should identify themselves only as 'Yugoslavs' to promote a single national identity. The use of the designation Serbs, Croats and Slovenes for political purposes was banned, infuriating the people. Croats and Slovenes distanced themselves from being identified as 'Yugoslavs' or Southern Slavs. Again, the Bosnian Muslims were not mentioned anywhere, which implied a denial of their existence. Inter-ethnic suspicions and a lack of trust were quietly brewing. Croats created an ultra-nationalist revolutionary party called the Ustaše to counter Serb domination, and the seeds of disintegration were thus sowed. King Alexander was assassinated in 1934 by extremist elements who felt aggrieved by King Alexander's decision to rename the country. The crown was passed to his 11-year-old son Peter II. Peter was young of age. Therefore, King Alexander's cousin Paul ruled as Regent to the prince until 1941, when Peter II finally took control as the King. Peter II's rule was cut short due to the start of World War II. Just before the Germans invaded Yugoslavia, the royal family escaped to London and setup a 'Government in Exile'.

During World War II, German forces occupied large parts of Yugoslavia. They created an 'Independent State of Croatia,' a neo-Nazi satellite state ruled by the infamous Croat Militia Ustaše. This newly created neo-Nazi satellite state of Croatia, dependent on Germany's support, lasted only until World War II. During their short rule, the Croatian militia Ustaše caused atrocities to thousands of people of other ethnic groups. The short rule by the neo-Nazi satellite state of Croatia during World War II sowed fresh seeds of distrust and hate, especially between the Serbs and Croats. During my stay in Bihac 1994-95, I heard many times from the local Bosnian population as they compared the Serb atrocities to those of Ustaše with deep hate, which reflected the hate embedded in history.

During World War II, Yugoslav nationalist forces fighting the Germans were split into two groups. Marshall Tito led the Partisans, the larger communist-led force in Yugoslav. Followers of the abdicated King living in an asylum in London were called the Chetniks. Yugoslav Partisans led by Marshall Tito were more successful in the resistance fight against the Germans. Partisans received recognition and support from the Allied Powers throughout the war, including allied airdrop of weapons, ammunition, and radio equipment. The radio equipment helped the partisans to coordinate their war efforts better. During World War II, Yugoslav partisans waged a very successful internal war of resistance against the German and other Axis Forces in the Balkans. After World War II, with the Germans defeated, Yugoslavia was again re-established as an independent state and Marshal Tito, the wartime leader, became the President. Yugoslavia was re-established as an independent country on the military strength of the partisan Army, where the Serbs dominated and held positions in the higher ranks.

President Tito of Yugoslavia had grown from the grass root level of politics to the highest position and was therefore aware of inter-ethnic hatred that could undermine the integrity of the newborn nation. President Tito was a Croatian and a successful wartime leader. He remained unchallenged and thus prevailed in suppressing inter-ethnic differences. He carefully balanced the inter-ethnic relationship. However, his policies appeared to be more aligned with the Belgrade agenda of the Serbs. This resulted in Serbs holding most of the dominant positions in the political and military hierarchy. The Yugoslav military had more representation from the Serb community and had effectively become the policing tool for Belgrade. In former Yugoslavia, the country's rule was centered around the single personality of President Tito, seated in Belgrade. As happens with most authoritarian regimes, Marshal Tito failed to create a politically acceptable successor or a political institution for continuity after his death.

After the death of President Tito in 1980, the Republics (*map below shows the former Yugoslav Republics*) that made up the country of Yugoslavia started raising voices for independence, possibly to break away from Serb-centric state control. Slovenia, having borders with Austria on the north, was first to break away in June 1991 with the least opposition from the Yugoslav Army. The war in Slovenia lasted for only ten days. Serb presence in Slovenia was small and insignificant; thus, the Yugoslav Army had no cause to fight and nothing to defend. However, the declaration of independence by Croatia resulted in severe fighting and bloody war due to the presence of a large Serb population in the southeast and in southern parts of the country.

Copyright: GEO CONSULTS, Bangladesh

The declaration of independence by Bosnia Herzegovina in May 1992 met a fierce onslaught from the heavily armed Serb Militia and units of the former Yugoslav Army. Bosnia Herzegovina had a large Serb population spread all over the country and a long border with Serbia on the east. It was difficult for the Bosnian Serbs to see Bosnia Herzegovina breaking away from Mother Serbia. Having seen the other republics that made up Yugoslavia gone, the long cherished Serb dream of the 'Country of Southern Slavs' was disappearing. The Bosnian Serbs and Serbia wanted Bosnia to merge with Serbia to form greater Serbia. Serb political leaders spewed hate speeches almost daily, turning the Serb community into hate frenzied killing machines. Slobodan Milošević, *President of Serbia* and Radovan Karadžić, President of Republika Srpska (Bosnian Serbs) were key figures in inciting the Serbs to kill and cleanse the non-Serbs. Both were later tried and convicted of genocide and war crimes. Balkan history does not support the notion of establishing a unified state integrating the different ethnic communities across the Balkans. Bosnia Herzegovina had three large ethnic communities the Bosnian Muslims, the Bosnian Serbs, who are Orthodox Christians and the Bosnian Croats, who are mostly Catholic Christians. By early 1900 a significant part of the population in Bosnia Herzegovina was Muslim.

War in Bosnia Herzegovina

The Declaration of independence from Yugoslavia by Bosnia Herzegovina on March 3rd, 1992, triggered the war in Bosnia Herzegovina. This declaration of independence was a follow up to earlier declaration of independence by Yugoslav republics of Slovenia and Croatia. In a 1991 census the population of Bosnia Herzegovina was 44 percent Bosnian Muslims, 31 percent Serbs *(Orthodox Christians)*, 17 percent Croats *(Catholic Christians)* and 8 percent other minorities. Serbs, who represented a significant part of the population of Bosnia-Herzegovina rejected the declaration of independence because the Serbs wanted to stay under united Serbia. Bosnian Serbs called upon their ethnic brothers in Serbia to support their fight to resist the move by Bosnia-Herzegovina to break away as an independent country. USA and countries in the European Union were quick to give recognition to Bosnia-Herzegovina as an independent country. However, among the Europeans a perception soon developed that a Muslim country in heartland Europe could in future promote jihadism in Europe and was therefore viewed as a security concern. Undermentioned dates reflect how much support the international community had initially for the Bosnia's independence, a euphoria which faded away in short time.

- March 3rd, 1992 Bosnia Herzegovina declares independence.
- April 6th, 1992, one month later, EU countries and USA recognize the new state of Bosnia Herzegovina.
- May 22nd, 1992, two months after declaration of independence Bosnia Herzegovina admitted into United Nations as a member state.

In Bosnia-Herzegovina the Muslim and Croat soldiers of the Yugoslav army broke ranks from the Yugoslav Army and combined with hundreds of volunteers from the local population to form the fighting brigades to defend their community against the Serb aggression. At the start of the war these forces were lightly armed and therefore took heavy casualties while fighting the heavily armed Serb forces who basically represented the former Yugoslav Army (JNA). Serb forces had full access to weapons storage depots of the Yugoslav Army and were thus able to arm themselves with heavy weapons, tanks and access to an unlimited supply of ammunition. The brutal war that raged for over three years in Bosnia Herzegovina and southern parts of Croatia is a story of immense human sufferings, killing of men, women, children, rape, and ethnic cleansing all of it happening in the middle of modern-day civilized Europe. Serb hatred for the Croats and Muslims was drummed up by their leaders to a frenzy that they drove them to demolishing of Catholic churches and Muslim mosques all because of intolerance of another religion or community. Bosnian war is a sad story of crimes committed against fellow countrymen, against neighbors and friends by fellow Serbs who were driven by fanatic nationalist feeling and religious hatred brewed by their leaders like Radovan Karadzic, Ratko Mladic, and Slobodan Milosevic.

European political leaders failing to agree to common political-strategic goals for the Balkans refrained from taking any effective steps like the use of NATO air power to stop the war. Bosnian war genocide may be difficult to find in contemporary history. All communities were affected by the atrocities committed and forced ethnic cleansing. Croats and Muslims were driven out of their villages in Bosnia Herzegovina by the Serbs from areas that now make up The 'Republic of Srpska,' a sub-state territory in Bosnia Herzegovina held by Serbs. A classic example of Serb atrocities and ethnic cleansing are the killings and forcing people out from their homes in enclaves of Srebrenica, Zepa, Gorazde and from the villages south of Bihac. Soldiers from the Bangladesh Battalion were to witness some of the dirt of the war around Bihac.

With both the European union and the UN efforts failing to stop the killing and the ethnic cleansing, the international community agreed to deploy UN Peacekeepers. What most people did not know is that the UN Peacekeeping mission was launched in Bosnia without a UN Security Council Peacekeeping mandate simply because the big powers had conflicting interests and agendas for Bosnia Herzegovina. Initial deployment of Peacek eepers came from Europe only, primarily as EU and NATO efforts to stop the fighting. As the war situation continued to worsen, and the European Union found it difficult to provide additional military resources and therefore started looking for military units from outside the region.

International media flashing graphic images of the ongoing brutality and killings of the Bosnian Muslims resulted in strong public demand within the Islamic World to intervene. Street agitations were witnessed in most Muslim countries asking their governments to do whatever they could to stop the atrocities on in Bosnia. Bangladesh was no exception, both the political and non-political organizations held public demonstrations and debates to protest the killings and ethnic cleansing of Muslims in Bosnia. In some Muslim countries, people even volunteered to go and fight for the Muslims in Bosnia. There were reports that some volunteer fighters, referred to as Jihadists, from different Muslim countries did participate in the fighting in central Bosnia. Reports of such volunteer fighters did emerge in Bihac also particularly with the Buzim Brigade, but during my fourteen months I did not see any foreign fighters. Serbs would have been happy to parade any captured volunteer fighters from outside the region to show outside involvement, but no such evidence was ever presented. So, it would be reasonable to assume that the story of foreign fighters in Bosnia was not based on any visible evidence.

At a time when the situation on the ground started worsening by the day, the European Union and NATO considered the induction of military units from non-European countries including Muslim countries. Non-European countries that contributed peacekeepers were from Bangladesh, Malaysia, Pakistan and possibly a unit from Kenya. Not many countries were ready to commit their forces to harm's way in a distant and dangerous conflict zone with unsustainable logistical support for their soldiers. The war in Bosnia Herzegovina had already been going on for over two years, waged by combined forces of Bosnian Serbs, Serb militia forces from Serbia and elements of the former Yugoslav Army. Serb forces had an unhindered supply of heavy weapons and ammunition from the former Yugoslav Army which helped the Serbs to wage the vicious war targeting both Bosnian Muslims and Croatian Catholic Christians. Places of religious worship were particularly targeted by the Serbs. Typically, Serbs after capturing a new village or township would immediately assign demolition teams to demolish the mosque or the Catholic church in the area. Around Bihac, I have witnessed such destruction of villages and specifically one Croat Catholic Church destroyed in a village south of Bihac city. This supports the allegations that Serbs waged the same kind of destruction across the whole of Bosnia Herzegovina. It is estimated some 100,000 people were killed during the war. However, I feel the real figures may be even higher. Fighting in Bosnia Herzegovina stopped after NATO air power was unleashed around Sep—Oct 1995. I have no reservation in stating that if European leaders had agreed to use the NATO air power earlier, as possibly the Americans had desired, the Bosnian tragedy would not have occurred to the scale that the world witnessed. Maybe the massacre of Srebrenica would not have occurred. As the world witnessed, in the end, it was NATO air power which forced the Serbs to agree to a ceasefire in Oct 1995 and to come to the negotiating table. Subsequent talks resulted in the Dayton Peace Agreement, which came into effect on 31 Dec 1995. All through the war, the lead European nations had opposed the US plan for an air campaign which could have brought an end to the ongoing conflict much earlier. The later part of the book discusses why the lead European nations allowed the war to continue.

UNPROFOR Urgently Needs a Battalion to Replace French Battalion in Bihac

The UN presence in Bihac started in 1992 at the request of UNHCR (*United Nations High Commissioner for Refugees*) to provide armed escorts for UNHCR convoys delivering humanitarian relief, mostly food items, to the local population. France provided the first military unit for Bihac in October 1992. The French unit was a mechanized battalion with around 1300 soldiers and 100+ light armored vehicles that had its own integrated logistic support. The French Battalion established camps at Coralici, Velika Kladusa, and Bihac City. Additionally, the French Battalion established several Observation Posts (OPs) along the confrontation line to monitor the war activities. The French Army practiced changing over the battalion every six months with a fresh unit. The tour for a French battalion in Bosnia was for six months. Having maintained a presence in Bihac for two years, the French Government told the UN in 1993 that it would not continue its presence in Bihac. Apart from Bihac, the French Army also had a large presence in central Bosnia, particularly around Sarajevo, where the French suffered several fatalities, mostly due to Serb sniping from surrounding mountains. With the French likely to discontinue their presence in Bihac, the UN started looking for a military unit from elsewhere to replace the French. The UN sent the proposal to Bangladesh early in 1993 to evaluate the feasibility of getting a mechanized infantry battalion.

Bangladesh's Government accepted the UN proposal in 1993 to send a mechanized infantry battalion to Bosnia subject to the UN providing the required equipment, including the APCs (armoured personnel carrier). As a follow-up to the deployment proposal, I went with a team of officers from the Bangladesh Army to Bosnia to understand the operational needs. We arrived in Croatia and visited UNPF (*UN Peace Forces*) HQ in Zagreb for briefings. We visited Bihac also, where we spent time with the French discussing operations and logistics issues for sustaining a battalion of twelve hundred soldiers. I split my team into operations and logistics groups to get separate briefings from my French counterparts. I spent more time with the outgoing French Commanding Officer to understand the task and challenges, which was very helpful. During the visit, I also met with the UNPF Deputy Force Commander in Zagreb, a Canadian General, and among other things, he told me with emphasis, *'Salim fighting the cold should be the major task.'* People in the UN were more concerned about us than the Bangladesh Army itself because UNPROFOR realized that we were coming from the tropics and might not have adequate preparation for the Yugoslav winter. On the operational side, there were differences of opinion within UNPROFOR (*UN Protection Force*) as to whether it was the right decision to insert a battalion from a Muslim country into Bihac, an enclave with a Muslim majority population. Furthermore, the Serbs had opposed the deployment of the Bangladesh Battalion in Bihac, which became a big concern for UNPROFOR. Because of the concerns voiced, the UN cancelled the deployment of the Bangladesh Army in 1993. After the cancellation of the deployment proposal, the earmarked infantry battalion in Bangladesh *(6 East Bengal Regiment)* was re-assigned for a UN mission in Kuwait. Personally, I did not find the mission very challenging, and to the surprise of many, I opted out of the deployment.

From early 1994, the war situation in Bosnia Herzegovina started to worsen. Further, with the deadline for withdrawal of the French battalion approaching and the French being firm in not extending their presence in Bihac, pressure on the UNPROFOR military planners increased. I know that UNPROFOR tried other options, like the redeploying a battalion already deployed in central Bosnia to Bihac. However, the proposed redeployment could not be done because the host country refused to allow its military units to be sent to Bihac. For most of the troop-contributing countries, the war experience in Bosnia Herzegovina over the past two years showed that sustaining a military unit inside an enclave surrounded by hostile Serb forces presented extremely difficult operational and logistical challenges. As such, the troop-contributing countries were not ready to accept the redeployment of their national entities inside enclaves such as Bihac.

The countries certainly had vivid reminders on their mind of the tragic events experienced in other enclaves in central Bosnia, such as in Zepa, Srebrenica, Sarajevo, and Gorazde and the challenges the host governments had experienced. Besides, the troop-contributing countries had also witnessed UNPROFOR's inability to protect the peacekeepers when threatened.

UNPROFOR BHC CIVIL AFFAIRS BIHAC

IN VIEW OF A POSSIBLE DEPLOYMENT OF BANGLADESHI BATTALION (BANGBAT) IN BIHAC, IN REPLACEMENT OF FREBAT3, COPO HAS REQUESTED OUR VIEWS OF THE POLITICAL IMPLICATIONS OF THE DEPLOYMENT OF BANGBAT IN BIHAC. PLEASE INDICATE POSSIBLE CONSTRAINTS AND YOUR RECOMMENDATIONS.

Subject: Possible deployment of Bangbat in Bihac

Reference: ███████████████ fax of 15 June 1994 to you on a possible deployment here of a Bangladeshi Battalion.

SERB REACTION

1.   The arrival of a Muslim Battalion to a nearly 100 percent Muslim pocket, where Serb security interests are fully engaged, is certain to aggravate tensions with the Serbs. This point seems to have been grasped in the cases of Srebrenica, Zepa and Gorazde; similar common sense would apply to the Bihac pocket. We understand that the Serbs have been extremely sensitive not only over TURKBAT deployment but even on replacement of Canadians by Dutch in Srebrenica. Also, unlike JORBAT and TURKBAT deployment, the possible BANGBAT deployment presents a case where a Muslim battalion would be deployed in a completely Muslim pocket.

2.   The Bosnian Serbs may ultimately not make the necessary difference between BANGBAT and 5 Corps. Manning by UNPROFOR of a buffer zone may face a difficulty. In the worst scenario, BANGBAT could be perceived by the Serbs as another Muslim force and thereby become a target.

3.   With respect to Krajina Serbs, UNPROFOR operations in Bihac depend on routine and continuing access from Zagreb through Krajina.

4.   Due to above constraints, suggest that the matter be raised with the Serbs before deployment of the BANGBAT.

REACTION OF BIHAC REGION AUTHORITIES

5.   For obvious reasons, the reaction of regional leaders to a possible BANGBAT deployment is more apparent in the Abdic-controlled North, than in the Sarajevo-oriented South.

UN Civil Affairs, page 1

2

6.   For the **North**, any successor composed of predominantly Muslim troops raises concerns. It is feared that these troops will favor the more Islamic South in the internal conflict, and unnecessarily raise Serb anxiety (both in Krajina and Bosnia.) Recall that Abdic deputies strongly objected the deployment of JORBAT in neighbouring Sector North (UKRBAT currently mans a narrow strip adjacent to the pocket so as to insulate the Bosnian border from the Jordanians.)

7.   The highest ranking civilian authority in the Bihac-controlled **South**, BiH Minister Mirsad Veladzic, avoided comment on the possible deployment of BANGBAT, simply saying "it's not up to us but Sarajevo to decide". There is no reason to believe that the Bihac authorities would have special objections to BANGBAT.

REACTION OF LOCAL POPULATION

8.   Deployment of BANGBAT would represent at first sight a blow to the expectations created by the French experience here. The FREBAT, a highly mechanized and sizeable unit of 1300 men, has a dominant and well-functioning role in the Bihac pocket. FREBAT also carries the prestige of France and represents the hope of a European connection for a population that increasingly feels isolated, disadvantaged and neglected.

9.   The fact that the Bangladeshi troops share the same religion as almost 100% of the local population could advance relations, especially in the South. Bangladesh's own domestic experience in relief operations would hopefully be utilized to maintain the same level of performance as the French.

10. These reactions, on FREBAT as well as on BANGBAT, can be classified as subjective, and certainly not insurmountable.

RECOMMENDATION

11. BANGBAT's possible deployment poses serious questions about the perceived partiality affecting the battalion's operational freedom of movement and action. In short, it is recommended that the Bangladeshi's, which are well experienced in recent peace-keeping operations elsewhere, should be saved from the highly controversial role waiting them with UNPROFOR in the Bihac pocket.

12. Before making the final decision on the deployment it would be necessary to make an evaluation from the military angle, concerning e.g. the strength and equipping of the battalion. Also, the reaction of the Serbs should be clarified well in advance.

UN Civil Affairs, page 2

In central Bosnia, Serbs were taking UN Peacekeepers hostage at their will, with UNPROFOR failing to take any protective measures. Thus, UNPROFOR was in a difficult situation in 1994, failing to find a battalion to replace the French battalion in Bihac. With no commitment in sight, UNPROFOR decided to relook at the only offer they had on hand, to deploy a Bangladesh infantry battalion to replace the French in the Bihac pocket. In mid-1994, UNPROFOR had intense discussions on the renewed proposal to deploy Bangladesh Battalion in Bihac. Documents copied below clearly indicate recommendations within UNPROFOR did not support the proposal to deploy a battalion from a Muslim country in Bihac an enclave with Muslim majority population. In the end we witnessed UNPROFOR giving orders to deploy the Bangladesh Battalion in Bihac against all internal recommendations not to do so. Serbs were very unhappy with the UNPROFOR decision, and their displeasure is reflected in the Serb memo of Nov 1994, copied below. Note the comments of the Prime Minister of the Krajina Serbs (*Croatian Serbs*) indicating that supply convoys will be blocked. Krajina Serbs (Croatian Serbs) controlled all territory over the entry and exit routes into Bihac. So clearly, UNPROFOR had invited trouble for itself.

In 1994, the Bangladesh Government agreed to the renewed UN proposal to send a battalion to Bosnia. I received orders to take command of the battalion going to Bosnia Herzegovina. However, the UNPROFOR staff like the year before, again raised the flag in 1994, voicing their concern that the Serbs would not welcome the UNPROFOR decision to deploy a battalion from a Muslim country into a disputed Muslim enclave. Events later would show how true the projected concern was, as it resulted in untold difficulties for the UN Operations, and sufferings of the Bangladesh Battalion resulting from weeks of food and fuel supply blockade imposed by the Serbs.

MESSAGE

1. During a meeting with the 'RSK' prime minister, Mr. Mikelić, in Knin on 11 November 1994, the prime minister gave me his intentions towards UNPF Convoys to Bihać.

2. Mr. Mikelic announced that 'RSK' government is in the process of making a decision concerning UN Convoys into the Bihać pocket. The decision has not yet been taken but is likely to be as follows: Only UN convoys with humanitarian purposes will be allowed into the Bihac pocket. All these convoys, although allowed, will be searched thoroughly for ammunition. The reason being that on previous occasions convoys for ▇▇▇▇▇▇ had provided 5th Corps with ammunition. 'RSK' authorities did not trust the situation any longer after being cheated once.

3. I commented on the above, that for this reason UNPROFOR had created 'Bihać Special Command'. Mr. Mikelić reacted on this by saying that in spite of 'RSK' requests that the Bangladesh troops be deployed in some other area to avoid suspicion, UNPROFOR had deployed a Muslim unit in the Bihać pocket.

Best regards

Serb PM's reaction to BanBat Deployment

With the limited time available for the deployment, a rush of activities started. In June 1994, a team of officers from UNPROFOR (*UN Protection Force*) came to Bangladesh to discuss the structure, composition, and training for the proposed mechanized infantry battalion. Bangladesh's acceptance for the deployment was on condition that the UN would provide the APCs, winter uniforms and other major equipment. The other requirement was for the UN to arrange training of Bangladesh soldiers on the use of the incoming APCs— Armored Personnel Carriers and equipment before deployment in Bosnia Herzegovina. The major concern for me was the proposed short time period for induction and training of specialized equipment, such as the APC. Additionally, I was worried about our lack of experience and preparedness for operating in freezing European winter conditions. Given the departure dates for the French, I knew that the deployment dates could not be pushed back. Therefore, I focused on advancing the deployment timings to allow the soldiers to settle down and adapt to the cold before the winter sets-in.

During the visit of the UNPROFOR team to Bangladesh, we asked the team members about the potential deployment location for the Bangladesh Battalion. The UNPROFOR Operations Staff indicated that they were considering three possible areas: Gorazde and Mostar in central Bosnia, and Bihac in Northwest Bosnia. Knowing about the intense fighting in Gorazde and Mostar, I was relieved to hear that the Bangladesh Battalion would be deployed in Bihac, which we thought at the time would see less intense fighting. Unfortunately, this turned out to be wishful thinking. Shortly after our arrival in Bihac, the fighting between the Serbs and Bosnian Muslims escalated. Scheduled UN convoys bringing the last group of our soldiers, equipment, food, and other supplies were blocked. The Serbs had previously warned UNPROFOR not to deploy Bangladesh Battalion to Bihac, and they kept their word by imposing the blockade. The Serbs were punishing the UN Peacekeepers, including the Bangladesh Battalion, for their battlefield losses around Bihac.

In Bangladesh, the soldiers are exposed to mild winter with temperatures around 15 Celsius / 59 Fahrenheit, except for a few areas in the north where night temperature may come down to 8 Celsius / 46 Fahrenheit for a few days only. During the winters in Bangladesh, the army continues to wear the summer cotton uniform with a light winter jacket which is certainly not appropriate for the Yugoslav winter. The Bangladesh Army has no inventory of winter clothes or equipment like what would be needed for an European winter. Eventually, the solution came with an UN-assigned contractor to provide the winter clothing and around seventy APC BTR-70 of the former East German Army. These were surplus stocks of former Soviet Union military stored in the warehouse in Slovakia for some future Warsaw Pact war with NATO, which never came. The contractor was also assigned the contract to provide four weeks of basic training to Bangladesh APC crews in a training camp in Slovakia. I had already reached Zagreb, Croatia early September 1994, so I could visit the training camp in Slovakia. To me, being able to visit this training base was an opportunity to also have a rare glimpse of the former Soviet Union's preparations for war with NATO; fortunately for Europe, the war never came. The warehouses were pre-stocked with weapons, ammunition and equipment like generators and winter uniforms.

The Serb blockade of ten weeks also demonstrated the failure of UNPROFOR leadership to protect the Peacekeepers. In my judgement, the failure had much to do with the professional ability of UNPROFOR military leadership at the time, the inability to take correct decisions or maybe the decisions were being made in order to appease the political leadership in home countries. To my mind regional political game was in play with operational decision making. I noticed the UNPROFOR General who came to visit me during the crisis in Dec 1994 was more concerned about his TV cameramen, than the peacekeepers like me standing next to him; he hardly talked to me after getting off his helicopter. Obviously, he wanted good TV coverage on the evening news. I guess the Serbs had read into the minds of the UNPROFOR leadership and therefore kept harassing the blue helmeted Peacekeepers at will, especially all through 1994.

Why Bangladesh

Bangladesh is among the top troop-contributing countries in the world for UN Peace Keeping missions. The first contribution was in 1988 when a group of military officers were sent as UN military observers (UNMOs) to Iraq—Iran border. Bangladesh has the practice of rotating people every twelve months. So, in 1989 I was selected to replace the Team Leader on the Iran—Iraq observer mission. I had packed up and was all set to go when Iraq invaded Kuwait, and all international flights to Iraq stopped. The planned rotation of Bangladesh military observers to Iraq was cancelled, and so was my UN assignment. My opportunity for UN assignment came a few years later when Bangladesh accepted the UN proposal to send a Bangladesh Army contingent of about twelve hundred soldiers for peacekeeping in Bosnia. Bangladeshi soldiers would be alien to European culture and living conditions, and most importantly, were not equipped or clothed for European winter. At the time, many people wondered how Bangladesh soldiers would deal with the freezing winters in Europe. I wondered why I was being deputed to command a battalion when I was in a higher position in command of a brigade. I also understood well this would be one of the most challenging and difficult missions. Was I being put to the test where the probability of failure was high like it happened with the Bangladesh UN mission in Rwanda a few years earlier. This mission was expected to be challenging for us because the UN Force in Bosnia comprised mostly of NATO military units or former Soviet bloc countries, and in theatre, people's expectations would be high. I also wondered if the Bosnian

Serbs, Croats or Bosnian militia would have faith in our ability to bring peace. Surely locals will compare the Bangladeshis with the NATO military units, and in all likelihood, Bangladeshis would be seen as minnows. Our soldiers had never seen snow and were neither equipped nor trained for mechanized warfare, which was the need. Despite the structural inadequacies, we did take up the job and accomplished it over fourteen months, September 1994—November 1995. I leave it to the readers to judge whether it was fair and how we performed.

Deployment Plan

Meeting the UN's deployment deadlines was a significant challenge for us. We were given just over three months from the initial discussions in June 1994 to replace the French military unit in Bihac by mid-October 1994. This included shipping the military equipment from Bangladesh to Croatia, acquiring and training on APCs and other equipment in Slovakia, mobilizing all elements in Zagreb, Croatia, and moving soldiers and equipment to Bihac, along with necessary logistical support. The timeline of three months to equip, train and transport our equipment half the world across to Bosnia was a recipe for disaster. Moreover, due to the war, ships traveling to ports in Croatia required insurance coverage to cover the potential war damage, which was an additional challenge. International shipping was disrupted, as many shipping lines stopped going to Croatia due to the added costs and war risks.

The battalion was split into four groups for our move from Bangladesh to Zagreb, Croatia, which was our staging camp. The first group was the 'Advance Party' under the Commanding Officer (CO—author). This Group was scheduled to go immediately to the mission area to coordinate with the UN agencies the arrival of the battalion, shipping, and the details of the operational deployment. The job for the Advance Party was critical given the short time available for the deployment. Between the Advance Party arrival time in September and the battalion operational deployment in October, we had barely six weeks to negotiate and coordinate all activities. Much of the equipment was to be provided under UN contracts, and we had no firm delivery dates as yet from the UN contractors. Additionally, some items were missing from the contracted list and had to be added to the UN contracts for supply within our six-week wait time at UN Camp Pleso in Zagreb. UN contractors in Zagreb locally supplied items included over twelve hundred pairs of boots appropriate for snow conditions, some mine shoes, some night vision goggles and APC spares. An interesting item negotiated was the food item demand list. The UN supply chain in Zagreb, Croatia, was primarily geared to support units from European countries. So, the UN system took a stretch to find contractors to supply Bangladesh food items like rice, lentils (dal), spices, lots of chilies, and fish items as regular food items. However, UN contractors did a good job in quickly sourcing the required items.

The second group was the 'Training and Logistics Group' of 316 officers and men sent to Slovakia around mid-August 1994. This training base in Slovakia was busy training units from other countries before us, so we could not be accommodated earlier. Our team in Slovakia received APC—BTR 70, ARV (Armored Recovery

Vehicle), dozens of trailer-mounted generators essential for forward field deployments and winter clothing uniforms of the former East German Army. After the unification of Germany, these East German Army uniforms were of no use to the German Army and possibly would have been thrown away. These uniforms came to good use for the Bangladesh soldiers during the freezing winters. This group spent about four weeks training in Slovakia on the newly acquired equipment, especially on driving and firing with the BTR 70 APCs. On completion of training, the entire group travelled by special train from Slovakia via Hungary to Zagreb, Croatia. In the mission area, the Bangladesh Battalion looked more like the former East German army soldiers. I did not like the look of East German uniforms, so I continued wearing my Bangladesh cotton uniform with thermal undergarments.

The third group was the 'Main Body' under the Battalion Second in Command (2IC) as the group in-charge. The group in-charge was responsible for organizing the Main Body into operational and support groups as per the new task organization. Soldiers were required to have medical inspections and vaccinations done and the documentation completed. This group was also responsible for collecting equipment like dozers and loaders from the Bangladesh Army depots, packing them and loading them on ships for carrying the items to Croatia.

The fourth group was the 'Rear Party' made up of people who would stay back in Bangladesh. They were responsible for taking care of all the stores and equipment of the battalion, including personal baggage, which was not to be carried to the mission area. The Rear Party also served as the battalion's rear link to the Army Headquarters in Dhaka.

The training and logistics groups consisting of APC drivers, APC gunners, APC Technicians, Field Engineers and Logistic Staff, reached and camped in Slovakia. They were housed in barracks where interestingly in the cold war era, Warsaw pact soldiers had camped for their annual military training. In 2022, the Bangladesh Army operates a few hundred BTR 80 APCs, which are far most advanced and easy to operate than their predecessor BTR 70 which were provided to us for use in Bosnia. BTR 70 represented the earlier Russian generation of APC technology. The BTRs 70 had twin engines meshed into a single gearbox, which was extremely difficult to calibrate or synchronize. Cold start of the BTR 70 APCs during the winter mornings remained a serious problem due to the poor quality of the batteries which would die if left idle through the winter night. As part of our operational SOPs, our APC drivers had to start and run the engines for about an hour at midnight to ensure that the batteries would hold up until the next morning.

I left for Croatia with the Advance Party, leaving the task of organizing the Main Body to the battalion second in command. I arrived in Zagreb on September 4th, 1994 with the Advance Party of 178 officers and soldiers. At this point, the battalion was split into three groups in three different countries. The Training and Logistics Group of 316 officers and men were training in Slovakia, the Advance Party of 178 officers and men were with me in Zagreb, Croatia and the Main Body of about 700 officers and men were in Bangladesh. While in Zagreb, I decided to visit Slovakia to see the progress on training and to coordinate the shipment of men and equipment from Slovakia to Zagreb by train. I was lucky to have the company of Major Lund, a very friendly and helpful Danish Army officer from HQ UNPF Croatia. We started from Zagreb in Croatia and drove through Budapest, Hungary, to Slovakia, which took us an entire day. The journey turned out to be an interesting experience for Major Lund and me. We were travelling in a white UN-marked car and alternating the driving between the two of us. From Croatia, once we arrived at the Hungarian border, we had to spend two hours at the border to get my visa which was finally issued. A few hours later, we arrived at the Slovakia border and had to wait more than one hour to get the entry visa. Two days later, on the return journey, when we were exiting Slovakia, I faced a similar wait at the Hungarian border. With an EU passport, the Danish officer did not have any issues with the crossings. However, despite my blue beret and UN ID, I had to undergo a time-consuming visa process while entering Hungary and again at the Slovakia border. The immigration officer at the border post had to send Faxes to national immigration control in their capital city and get approval because I was carrying a non-EU passport. Leaving apart the wait at the borders, the experience of the drive and especially of going through Budapest was nice and enjoyable.

The training camp in Slovakia was a very large and impressive training facility. My visit showed me how much the old Soviet Union had pre-stocked supplies they would need for a war with NATO that fortunately never came. We, as UN Peacekeepers, ended up using part of that equipment and the facilities. After completing training in Slovakia, our team was transported to Croatia by a special train loaded with APC BTR-70s, generators and a host of other equipment. The soldiers and stores arrived in Zagreb about mid-September of 1994 in two special train loads. Unloading the trains was an experience for us; unlike the time-consuming practice in Bangladesh of unloading one vehicle at a time to go over the ramp, the APCs and ARVs turned on the train carriage itself to drive onto the station platform, which was at the same level. After that, the vehicles, including the APCs, drove to the camp in small packets. I spent the day shuttling between the station and the camp as we took the whole day to unload and transfer the equipment to Camp Pleso near Zagreb airport. With two train loads, we received something like 68 * BTR-70 APCs, a few tracked ARVs, a large number of jeeps, trucks, trailer-mounted generators and other field equipment.

Stores from Slovakia also included loads of winter uniforms of the former East German Army. After issuing the uniforms the next day, the Bangladesh battalion suddenly turned from green camouflage to brown camouflage, resembling an East German Army unit. At the UN Camp in Zagreb, we made use of the available time by organizing training that included winter adaptation and survival techniques, situational awareness, and landmine awareness. All the warring groups in Bosnia had extensively used mines, so our soldiers had to be extremely cautious and watchful. APC drivers continued their driving practice in the limited space at UN Pleso Camp, where soldiers from all nationalities were housed in large tented accommodations called Rub Halls. Adapting to the new culture was an important part of the soldiers' training. Medics from the USA gave classes on winter adaptation and survival techniques. The training included some interesting tips, such as discontinuing the typical Bangladeshi practice of rubbing oil on the body after a bath because it spoils the skin's insulation. We also learned that drinking hot tea or coffee was not good for soldiers on outdoor night duty because caffeine intake results in frequent urination, which can lead to cold exposure for night sentries.

In the last week of September 1994, the Bangladesh battalion received operational orders from UNPROFOR to deploy into Bihac and relieve the French battalion. As the contingent commander, I was still concerned about the driving skills of our APC drivers and the ability of all our drivers, including those of soft vehicles, to operate in snowy conditions. In Bosnia, many fatalities were caused by vehicle accidents during winter, including the dreaded black ice on mountain roads. In early October 1994, just three weeks before deployment, we received the unsettling news that our equipment, shipped from Bangladesh, was still sitting at Alexandria port in Egypt, as the UN shipping agent had not been able to secure a ship for transshipment to Croatia. This included vital equipment such as the field hospital, workshop, and essentials for supporting the battalion. We later discovered that due to the ongoing war in Croatia, shipping companies were reluctant to take on the risk of transporting goods to Croatian ports and required special insurance for war zone coverage. This resulted in a delay of a month as the containers had to wait at the port in Egypt until a suitable ship was found for transportation to Croatia.

In meetings at HQ UNPF Zagreb Croatia, in the weeks preceding deployment, the issue of the failure of the UN shipping agent to deliver the battalion equipment was repeatedly raised and every time, the matter was pushed aside with the decision to pursue the matter with UN New York. The option to delay the deployment till the ship's arrival was discussed, but was not agreed upon by the UN Command. Certainly, the ship stores should have arrived at the destination port in Croatia by the first week of Oct 1994, which would have enabled the battalion to receive all its operational equipment before going into Bihac. Unfortunately, UNPROFOR, instead of accepting the failure of the UN shipping, started blaming the Bangladesh Battalion for arriving without its operational support stores. Failure of the UN shipping agent to deliver our stores resulted in multiple operational difficulties. Engineer specialist vehicles, heavy weapons like the anti-tank guided missiles (ATGM), battalion mortars, heavy machine guns (HMG) and the extra load of the battalion reserve ammo were not in hand when the battalion started its deployment in Oct 1994. The UN Chartered ship finally arrived a month after our deployment. To further added complication, there were difficulties in getting road movement approval from the Serbs. Therefore, the heavy weapons like the ATGM, Mortars, HMG, and reserve ammunition had to be airlifted later by UN chartered MI-26 heavy lift helicopters . UNPROFOR could not push back the deployment dates because the French were adamant about leaving by the given deadline. UNPROFOR's decision-making process is to be blamed more than the French, who had informed UNPROFOR almost a year back about their intention not to continue their presence in Bihac.

2007/8/Ops/10   26 October 1994

SHIP CARRYING CARGO FOR BANBAT-UNPROFOR

1.   It is intimated that the following message has been received
from the Commanding Officer of Bangladesh Battalion (BANBAT) in
UNPROFOR, which is self explanatory :

     Quote

     1.   SHIP CARRYING 19 CONTAINERS, SOME LOOSE LOAD AND 26 VEHICLES
     SAILED FROM BANGLADESH ON 17 SEP 94. IT WAS EXPECTED AT RELJEKA ON
     08 OCT 94. WE WERE LATER TOLD SHIP IS EXPECTED ON 20 OCT 94. WE
     NOW HEAR FROM SMOO THAT OUR SHIP LOAD IS STILL AWAITING TRANSHIP-
     MENT AT ALEXANDRIA (EGYPT). THIS LEAVES US IN COMPLETE STATE OF
     UNCERTAINTY.

     2.   PLEASE NOTE THAT TILL THE SHIP ARRIVES, BANBAT WILL BE WITHOUT
     FIELD HOSPITAL, FIELD WORKSHOP, ENGINEER ASSETS AND MANY SUPPORT
     UTILITIES. THIS PROBLEM IS IN ADDITION TO THE FACT THAT CURRENT
     DEPLOYMENT IS WITHOUT SPARES FOR BTRS, OTHER GERMAN VEHICLES AND
     THAT THE GERMAN COMPANY HAS FAILED THE UN CONTRACT FOR DELIVERY
     OF SLEEPING BAGS TILL TODAY.

     3.   REQUEST TREAT THIS AS URGENT AND TAKE APPROPRIATE MEASURES.
     REGARDS.

                                              Unquote

2.   In view of above, you are requested to kindly take up the matter
urgently with UNOPRO so that the problem is resolved immediately.

3.   Your cooperation is much solicited. Warm regards.

                         ABDUL HAFIZ
                         Lieutenant Colonel
                         for Principal Staff Officer

Distribution :

External:

Action :

Bangladoot, New York   - BY FAX
Attention : Mr. Mohammad Ziauddin
            Counsellor
                         1

Copy of letter sent by Bangladesh Government to UN HQ, New York, on shipment delay

Deployment Timeline

The deployment dateline below helps better understand the constraints under which the Bangladesh battalion was deployed in Bihac, Bosnia Herzegovina.

Preparation

- **June 1994**: The decision to deploy Bangladesh Battalion to Bosnia is confirmed. UNPROFOR Team arrived in Bangladesh to finalize the deployment plan. It confirmed that the Bangladesh battalion would be replacing the French Battalion in Bihac, who were due to pull out by the end of October of 1994.
- **July 10th, 1994**: As Commanding Officer (CO) of the battalion I joined 10 East Bengal in Savar, Bangladesh, the core element of the Bangladesh Battalion for Bosnia. Other support elements also joined to form the restructured Bangladesh Contingent for Bosnia mission.
- **Mid August 1994**: 'Training and Logistics Group' of 316 Officers and Soldiers leave for Slovakia to receive the operational equipment provided by the UN and for the training of APC crews.
- **September 4th, 1994**: An advance party of 178 personnel, with the commanding officer, arrived in Zagreb, Croatia.
- **September 10th, 1994**: On 10 Sep CO Bangladesh Battalion (author) and Col Lund, G3 plans from HQ UNPF, visit Slovakia to see the training arrangements.
- **Mid-September 1994**: Main body of about 700 personnel from Bangladesh arrived in Zagreb,

Croatia, on two UN chartered flights. The battalion is camped at the UN Logistical Base Pleso in Zagreb, Croatia, preparing for deployment to Bihac.

- **End September 1994**: Training and Logistics Group of 316 Officers and Soldiers, after completion of 4 weeks training in Slovakia, arrive in Zagreb, Croatia on special train transiting via Hungary.
- **End September 1994**: Bangladesh battalion receives UNPROFOR operational orders for deployment to Bihac to relieve the French battalion. UN chartered ship fails to arrive to deliver the battalion equipment. This is repeatedly mentioned in weekly operational meetings in Zagreb, but the UN Logistics Department in New York did not appear to be seriously concerned about this delay and the likely impact it would have on Bangladesh Battalion operational capability.

Deployment into Bihac Enclave

- **October 18th, 1994**: The first convoy of the battalion deploys in Bihac. The French Battalion starts thinning out to make space for the Bangladesh battalion. Briefings and joint patrols are conducted in forward areas, particularly covering the OPs (*observation posts*) on the confrontation line. OPs were manned 24/7 to observe and report on incursions, movement of armed groups, shelling, and firing, particularly when civilians were targeted. Two more convoys of the Bangladesh battalion arrive in Bihac with additional soldiers, arms, and ammunition.

- **October 21st, 1994**: Bangladesh battalion takes over the operational responsibility, and the last elements of the French battalion leave Bihac.
- **October 25th, 1994**: Bosnian Army 5 Corps surprises everyone by launching a major offensive in the south capturing large areas, possibly taking advantage of the transition period between the Bangladesh battalion and the French battalion.
- **October 28th, 1994**: The First UN Logistical convoy with food and fuel for Bihac fails to arrive as Serbs impose a total blockade for UN convoys to show their anger for the Bosnian 5 Corps attack.

Tasks of Bangladesh Battalion

1. Provide assistance to UNHCR to deliver humanitarian relief. Provide convoy protection when requested—UN SCR 776.
2. Provide protection to other humanitarian agencies when requested with the approval of UNHCR—UN SCR 776.
3. Protect UNHCR storage facilities when requested— UN SCR 776.
4. Provide convoy protection for convoys of released detainees when requested by ICRC and on approval of HQ UNPROFOR—UN SCR 776.
5. Deter attacks on 'Safe Areas.' Monitor ceasefire in Safe Area; Bihac was UN declared Safe Area. Promote withdrawal of military and para-military units other than those of the Bosnian Government from Safe Area—UN SCR 836.
6. Occupy key locations on the ground to monitor and report on firing incidents—UN SCR 836.

After receiving the UNPROFOR final orders to deploy to Bihac, we organized the battalion into four convoy groups for travel from Zagreb to Bihac. As Commanding Officer, I started for Bihac with the first convoy on September 21st, 1994, leaving Zagreb in the early morning, probably around 7 am. Our convoy was a mix of APCs and trucks carrying stores and soldiers. We left Zagreb early because of the long journey ahead and anticipated delays. My major concern was the inadequate driving skills of our APC drivers, who had converted from truck drivers to APC drivers only a few weeks back and received just four weeks of basic driving training in Slovakia. The fact that there was approximately 100 km of road journey ahead with the greater part of the journey through the conflict zone created additional mental stress on us, especially the drivers.

We began our journey under the escort of a UN Military Police (MP) team from Denmark. However, within the first few kilometers of driving through the city roads of Zagreb, one of our APCs accidentally ran over a civilian car at a road intersection. Fortunately, the driver of the car survived and the UN MP team took care of the incident. This incident highlighted my concerns about the skills of our APC drivers and their lack of experience. The mental pressure of driving into a war zone only compounded the issue. After the incident, our drivers became even more cautious and nervous. Despite the setback, we continued our journey, moving at a snail's pace and taking nearly eight hours to complete the 100 km journey to Bihac.

French soldiers gradually started leaving Bihac as the first group of Bangladesh soldiers arrived, which helped release pressure on accommodations, food, and fuel stocks. Over the following weeks, two more convoys reached Bihac without any further accidents. By October 18th, 1994, three of the four planned convoys of the Bangladesh Battalion had arrived in Bihac with soldiers individual weapons and essential equipment. We were spread in three camps. The camp accommodations consisted of prefabricated containers, with each camp generating its own electric power for heating and lighting, which was essential during the winters. The Bangladesh Battalion officially took over operational responsibility of Bihac from the French Battalion on October 21st, 1994 as the last elements of the French Battalion departed.

A handover ceremony was arranged with a small contingent of the French and the Bangladesh soldiers paying tribute to the French flag as it was lowered and to the Bangladesh flag as it was raised at the Bihac Corallici Battalion HQ location. The last elements of the French battalion left Bihac on October 21st, 1994.

Handover Flag Ceremony. CO Bangladesh Battalion (author on left), and outgoing CO French Battalion, 1994 (Source: Author's own records)

In the first weeks of the Bangladesh Battalion's deployment, soldiers with their personal weapons and pouch ammunition had arrived in Bihac. The logistical element soldiers, heavy weapons, equipment, and reserve ammunition were supposed to arrive in the 4th and last convoy scheduled to arrive on October 28th, 1994, along with the first UN logistical convoy carrying food and fuel. Unfortunately, none of the two convoys were unable to reach Bihac due to the outbreak of fighting between Serb and Bosnian Muslim forces in Bihac.

On October 25th, 1994, just three days after the departure of the French Battalion, the Bosnian Army 5 Corps launched a major offensive and successfully broke out of Bihac in the south. This was the first time in over two years of war that the Serbs suffered a humiliating defeat. The Bosnian Army 5 Corps captured the dominating Grabez plateau and vast areas further south up to Ripac. Over the years, Serbs had used the position at Grabez plateau to snipe at civilian movements in Bihac city. The Bosnian Army 5 Corps continued its attack and captured Bosna Krupa, an important city along the Una River and close to Bihac.

The battlefield scenario had changed dramatically just a week after the Bangladesh Battalion's arrival, so the Serbs accused the Bangladesh Battalion of having assisted the Bosnian Army 5 Corps. However, the accusation does not seem logical as the Bosnian Army 5 Corps attack was a major operation that must have been planned months before. The timing of the attack was brilliant, taking advantage of the transition between the French and Bangladesh militaries, achieving the element of surprise, a feature behind the success of most operations in military history. I also feel the Serbs had become nonchalant over the years, having defended the dominating Grabez plateau against multiple attacks in the past. It was later learned that the Serbs had reduced the number of soldiers in these positions in recent months to fight in central Bosnia, an important information which must have reached the Bosnian Army 5 Corps as an intelligence input.

Deployment Challenges

For me, the deployment timeline for the Bangladesh Battalion presented a significant challenge as most of the essential activities for deployment were beyond my control. One of the biggest challenges was that the Bangladesh Army did not have a mechanized infantry unit in June 1994, yet they agreed to deploy a mechanized infantry unit on a notice time of three months. I still wonder if people understood the implications of their decision. At the time, the only visible APCs in the army were a few Russian APCs that the Bangladesh military units had brought with them from Iraq while returning home after the 1991 Gulf War. A much more significant challenge ahead of us was the European winter, which demanded special clothing, equipment, and winter survival training, none of which were present.

I clearly remember one of my colleagues from staff college telling me about his meeting with the British High Commissioner in Dhaka, who commented that even the British Army, with their experience and fully equipped mechanized infantry battalion, would find it challenging to commit deployment of a mechanized unit overseas to a conflict zone in such a short time as asked by the UN. The British High Commissioner in Dhaka had rightfully questioned the wisdom of such a decision. Mechanized operations require more than simply APC driving skills.

APC crews are expected to operate under fire and respond appropriately. Operational training enables the crew to do fire and move drills to protect the vehicle and allow the gunner to respond. In 1994, the Bangladesh army did not have APC drivers and APC Gunners with any operational experience and training. Truck drivers were selected and assigned the role of becoming APC drivers. The inadequate skills of our APC drivers were demonstrated on our very first journey into Bihac when the lead APC crushed a civilian car within the first hour of the journey.

In the military, we are trained not to question, so I moved on to make the best of what I could. As I write this book in 2022, about 25 years after the event, I know that the Bangladesh Army of today has come of age and is more comfortable with such deployments.

As off 2022, Bangladesh Army now operates hundreds of APCs and other equipment deployed in various UN missions overseas, but this was not the case in 1994 . Not only that, the needed equipment was non-existent at the time, but to my frustration, I had great difficulty in making people understand what was needed, even for simple things like a soldier's sleeping bag. This was the first time army was buying 'bulletproof jackets with added ceramic plates', and being a weapons instructor; I personally conducted the acceptance quality for the bulletproof jackets by conducting a live fire test at Bangladesh Ordnance Factory premises. This test was carried out in the presence of the bidding suppliers. I can still remember only three of the ten test jacket samples withstood the live fire test from five meters. Suppliers who saw their sample jackets pierced walked away without complaint and in embarrassment. Tests were kept transparent for every competing supplier and official to see; as a result, I did not receive any request to favor any particular supplier. This was my way of dealing with the procurement lobby which proved to be quite effective.

Another example was my request for global positioning system (GPS). In 1994 Bangladesh Army did not have GPS, which in the context of operations in Bosnia was essential to get the target coordinates for any request for NATO air support. Getting just a few GPS became a difficult issue, and I was never given GPS by the army. Today even cheap smartphones have embedded GPS, so people now may find it difficult to even perceive the challenges we had to face .

On the operation side, the big challenge was planning and executing every convoy journey from our UN Support Base in Zagreb, Croatia travelling through the Serb-held territory into Bihac, a journey of approximately 100 kms. For each journey, the convoy manifest with the vehicle registration plate number and names of soldiers specific for each vehicle had to be submitted a week before the scheduled move day for travel approval from HQ RSK (*Headquarters of the Republic of Serb Krajina—Croatian Serb Military located at Knin. RSK controlled our move corridor areas which were part of their self-declared republic in southern Croatia bordering Bosnia*. On the day of the journey, we had to ensure that the soldiers were seated as per the vehicle manifest approved by the Serbs. Interestingly these areas held by Croatian Serbs were called UN Protected areas (UNPA), but the UN had no control except for reporting incidents. There were four such areas, as shown on the map on page 191, Sector North, Sector South, Sector West and Sector East.

The UN Force in UNPAs were a mix of unarmed UN observers and armed military units or battalions, mostly from Europe. UN forces had no real control over what the Serbs were doing except reporting incidents. Within the UNPAs, at Serb checkpoints, UN personnel were harassed for silly reasons on the excuse of checking. At the checkpoint, if they found a person sitting at the back of the truck, who should have been sitting next to the driver as per the manifest submitted earlier, they would not allow the vehicle to proceed. Names and ID numbers of the UN personnel were expected to match the vehicle number during checking for onward travel clearance. At the Serb check post, nobody was allowed to get off the vehicle. This was a journey of approximately 100 km from Zagreb and would normally take 6/8 hours. But in the worst case, the journey could be overnight stoppage on the roadside if the Serb soldier at the checkpoints did not clear the onward move. I recollect once one of our convoys did wait overnight at a Serb checkpoint with soldiers sitting in trucks all through the winter night; it must have been a torturous, freezing night for our soldiers. I noted that the Serb checkpoints were mostly placed in a way to stop the incoming vehicle on the upslope. The vehicles had

BanBat truck accident. Fortunately, no human casualties, 1995
(Source: Author's own records)

manual gears, and it was not easy for drivers to stop on an upslope and move again. From a military point of view, the placement of checkpoints to stop vehicles upslope may have had a good reason because the APC and other vehicles would be exposing the soft belly or underside to the Serb tank looking down from the checkpoint.

Driving under snow conditions, especially with black ice in mountainous areas, caused many casualties for Peacekeepers in Bosnia Herzegovina. It was a big challenge for Bangladesh Battalion as our drivers had never seen snow and found it difficult to comprehend how snow conditions can cause vehicles to skid or go out of control. With repeated brainwashing on the subject, we made our drivers extra cautious. My instructions were to drive in low gear and avoid using the brakes. I understood well that the vehicle would skid out of control if the brakes were used on a surface covered with black ice. Fortunately, we had only a few snow-related vehicle accidents, but no fatalities. The picture below shows the first skid accident, fortunately, without fatalities. Note the recovery truck had to park off the road to get a firm grip and avoid skidding because the road surface was covered with black ice.

Controversies on Bangladesh Battalion Deployment in Bihac

Deployment of the Bangladesh Battalion to Bihac came with controversies questioning our neutrality as Peacekeepers. Such controversies resulted from two factors. First, within days after the arrival of the Bangladesh Battalion, there was a sudden surge in fighting. Bosnian

Army 5 Corps in Bihac had launched a successful break-out attack only three days after the arrival of the Bangladesh battalion. The second likely factor that contributed to the controversy was the change in the battlefield status quo since the start of the conflict. For the first time, Serbs were on the run and had lost a lot of ground south of Bihac. Serbs had also lost many heavy weapons, including artillery guns and a few tanks, as they retreated.

I could sense from a series of activities within UN circles that UNPROFOR HQ at Sarajevo appeared to be perturbed to see the Serbs losing ground and heavy weapons. This change in battlefield status quo was most likely not in line with the larger strategy in the minds of key players. Throughout the war until Oct 1994, people were used to hearing the war news of Serb gains and the Bosnian Army taking the beating. This sudden reversal on the battlefield left people wondering about the reasons for the change on the battlefield. The only new factor they could find was the arrival of the Bangladesh Battalion in Bihac. Regrettably, some UNPROFOR officers in the senior and command positions also started suspecting that the arrival of the Bangladesh Battalion in Bihac may be the contributing factor to the battlefield changed scenario.

So, the UNPROFOR ghost hunt began to find evidence of the lack of neutrality by Bangladesh Peacekeepers. We have reasons to believe that UNPROFOR HQ was giving verbal instructions to UNMOs *(UN Military observers)* in Bihac to keep an eye on the Bangladesh Battalion. I suspect the instructions were verbal because nobody wants to leave behind a documented trail on controversial issues. Some people in UNPROFOR also re-raised the old issue, questioning how the Bangladesh Battalion was deployed in Bihac in spite of staff recommendations against it.

In Bihac, the UN presence was of two kinds, the Bangladesh Battalion and the UNMOs *(UN Military observers)*. UNMOs operated independently, reporting to CMO (Chief Military Observer) at Sarajevo under HQ UNPROFOR. In weeks after our deployment, I witnessed a sudden change in UNMO positioning in Bihac. Egyptian UNMO was taken out of Bihac; I suspect being a Muslim, he was no longer trusted as neutral. I believe that other UNMOs were tasked to keep watch on the Bangladesh Battalion and report on our activities. I say so because I suddenly noticed a significant change in our relationship. UNMOs started visiting our camps frequently and asking odd questions to Bangladesh officers about our operational activities. It is not under UNMO's mandate to spy on other UN Peacekeepers. My response was to immediately restrict the UNMO movement inside my camp premises, including the stoppage of free meals that the UNMOs enjoyed in our officer's mess. To me, courtesy cannot be a one-way affair, and we had no obligation to feed the UNMOs for free. I know that under the UN system, the UNMOs were salaried generously to pay for their accommodation and meals. For their operational needs, they would drive around in a UN provided car or SUV and refuel from any nearby UN facility like the Bangladesh Battalion camp. However, as the situation in the Bihac enclave deteriorated for safety reasons, I allowed some UN staff, like the UNMOs and UNHCR Chief, to be accommodated inside our battalion camp, and we also provided them with meal support.

Wartime Bihac

Bihac is an enclave with a Muslim-majority population on the northwest tip of Bosnia, bordering Croatia. During the war, Bihac remained surrounded by the Bosnian Serb Army (*BSA—the self-declared Republic of Srpska in Bosnia*) and the Croatian Serb Army (*RSK—self-declared Republic of Serb Krajina in southern Croatia bordering Bosnia on NW*). As a result, Bihac remained under siege for over three years until the end of the war in 1995. Bihac's wartime population was around 180,000, with the majority being Muslims. Most of the Serbs living in Bihac had left early in the conflict, leaving only about 300 Serbs still in the area in 1994. Bihac also had a Croat population of around 5000 people, mostly living in Bihac city and Croat villages located south of Bihac city. The encirclement of Bihac by the Serb military forces caused immense suffering for the inhabitants due to an acute shortage of food, fuel and other essentials including electricity. Before the war, electricity for Bihac came from the Yugoslav national grid, which was now under the Serbs' control. To strangulate the Bihac population, Serbs disconnected electricity into Bihac. Forests surrounding the villages started disappearing as people stocked wood for cooking and heating. Fuel needed for the few essential vehicles and generators was smuggled in and sold at a high price. In 1994/95, the price of one liter of petrol was 12-15DM German. Smuggling rackets run by the Serbs brought fuel and food into the pocket. Serbs often prevented the UNHCR from bringing in food convoys on which the local population depended. During the war period the payable currency in the Bihac was German DM.

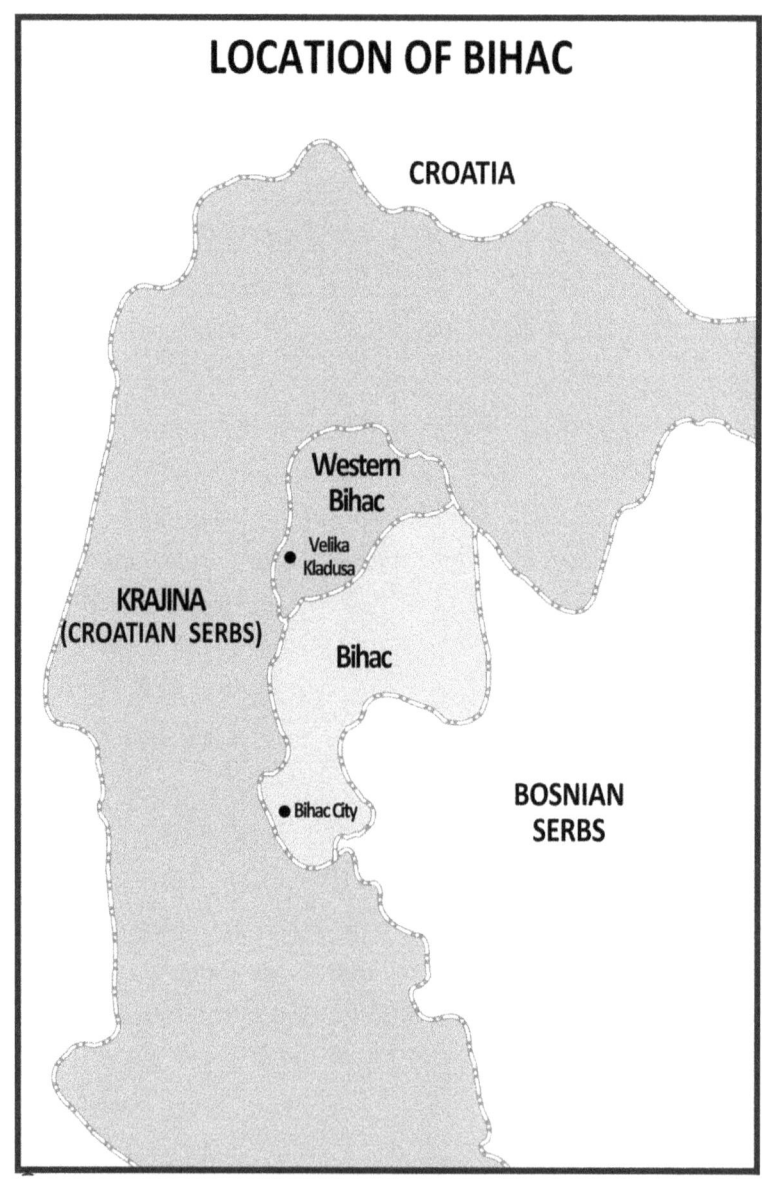

Copyright: GEO CONSULTS, Bangladesh

BIHAC , BOSNIA HERZEGOVINA COMMODITY PRICES 1994			
Salt	40 DM / kg		$23
Sugar	50 DM / kg	Approx price in USD in 1994 conversion rates. 1 USD = DM 1.742	$29
Cooking Oil	40 DM / Litre		$23
Flour	1,000 DM / Bag		$574
Coffee	60 DM / kg		$34
Petrol	15 DM / Litre		$9

Commodity Prices, Bihac, 1994

Serb policy of strangulation of Bihac possibly had two objectives. First, the shortage of food and essential supplies created enormous pressure on the population which could weaken the will of Bihac people to continue the resistance fight. Families with young children to feed were most affected. Next, the Serb enormously benefitted from the blockade, which kept smuggling alive. Krajina Serbs operated smuggling certainly with the help of insiders from Bihac. The table above reflects the average prices of essentials in German DM as 1994/95 and in USD–$ with the exchange values prevailing at the time. The black market thus became an essential element of the war and the wartime black economy for the Serbs. Where were people getting the money to buy the essentials at such exorbitant cost? Locals told me that most people had relations or family members working in Germany, Austria or other European countries and were sending money on non-banking channels. Many of the payments were actually done in Germany and Austria.

Bosnians in Bihac had formed their resistance army under an ex-Yugoslav military officer named Atif Dudakovic.

The new military formation was named the Bosnian Army 5 Corps. Incidentally, the former Yugoslav Army military formation in Bihac also had the same name as 5 Corps. As part of the Bosnian muslim resistance force in July 1992, the first Muslim brigade and the Croat brigade were formed in Bihac. In August 1992, the Cazin Brigade and the Velika Kladusa Brigades were formed; in September 1992, the Krupa Brigade and Buzim Brigade were formed. These newly established outfits comprised mostly civilian volunteers with rudimentary military training and some ex-Yugoslav military personnel. These forces initially carried only light weapons and relied more on weapons they captured later from the Serbs. In the southern part of Bihac, several villages existed with a fairly large Croat population, mostly made up of Catholic Christians. Serbs are Orthodox Christians and that was the religious divide between Serbs and Croats. Croats in Bihac formed a Croat Brigade. This Croatian Brigade (*Croatian HVO unit*) of around 1200+ soldiers operated under the Bosnian Army 5 Corps as an autonomous unit of the Army of the Republic of Bosnia Herzegovina. So, in Bihac, the Bosnian Muslims and Bosnian Croats formed a joint resistance force. In 1993, the Bosnian 5 Corps had eight fighting brigades.

The document copied below is an extract from my report that I had sent to UNPROFOR (United Nations Protection Force) on my assessment of the Bosnian Army 5 Corps in 1994. Three of the brigades, which had soldiers from the Bihac northern area, later defected to join and fight for Fikret Abdic, a Muslim rebel leader from Northwestern Bihac. As a result, Bosnian government forces, the Bosnian Army 5 Corps, were now left with five fighting brigades.

Banbat Assessment of Bosnian Army 5 Corp in 1994

a.   BiH (5th Corps).

(1)   Str. This Corps consists of 8 Bdes with its HQ at Bihac town. Str of this corps is approx 20,000 inf men. Bdes are 501, 502, 503, 505, 506, 510, 511 and 517. 101st HVO Bde is also under its orbat as a special one which consists of croats.

(2)   Comds/Loc of HQs.

(a)	5th Corps Comd	–	Gen Atif Dudakovic HQ at Bihac Town
(b)	501 Bde Comd	–	Brig Sarganovic HQ at Bihac Town
(c)	502 Bde Comd	–	Brig Abdic Hamdija HQ at Bihac Town
(d)	503 Bde Comd	–	Brig Delalic HQ at Cazin
(e)	505 Bde Comd	–	Maj Jusic Sead HQ at Buzim
(f)	506 Bde Comd	–	Brig Nirlikovic Miad HQ loc in between Vrnograc and Buzim
(g)	510 Liberation Bde Comd	–	Col Amir Avdic HQ at Cazin
(h)	511 Bde Comd	–	Brig Sedic Mirsad HQ at Pistaline
(j)	517 Bde Comd	–	Brig Nadarevic Ibrahim HQ at Pjanici
(k)	101st HVO Bde Comd	–	Col Ivan Prsa HQ at Bihac Town

(3)   Arms & Eqpt. The 5th Corps is strong in Inf but lacks hy wpns. However they have 6 X T-55 tks(captured from serbs), 1 X Praga (30mm Gun), some 20mm AA Gun, few 76 ZIS-3, some atk guns, 1 X T-34 Tk but it is out of order (captured from Abdic Forces). AAA is used in the area as MGs against grd tgts.

b.   BSA.

(1)   Str. It consists of two Corps who are deployed or the South and South Eastern side of the pocket.

BanBat Assessment of Bosnian Army 5 Corps, 1994

So the Bosnian muslim forces in Bihac were fighting both the rebel Abdic forces and also the Serb forces. The seizure of Bihac ended in August 1995 after joint operations by the Bosnian Army 5 Corps and the Croatian Army. Bosnian Army 5 Corps in Bihac attacked the Krajina Serbs (*Croatian Serbs*) from the south as the Krajina Serbs were busy fighting the main assault of the Croatian forces on the north and north-west. Sandwiched between the Croatian Army on the west and the Bosnian Army 5 Corps on the east, the Krajina Serb forces (*Croatian Serbs*) buckled and were defeated quickly. Bihac was liberated from Serb strangulation after three long years. Krajina Serbs discarded their weapons and fled as refugees to Serbia or neighboring Serb-held areas in Bosnia.

UNPROFOR Commander General Rose's first and only visit to Bihac, Dec. 1994 (Source: Author's own records)

Author briefs UNPROFOR Commander General Rose during his visit to Bihac, Dec. 1994 (Source: Author's own records)

During the Cold War, Bihac was an important military location for the Yugoslav military. The former Yugoslav military had built one of its largest airfields in Bihac with multiple runways. The taxiway extended into tunnels dug into the mountains where the aircraft could be hidden. The entrance to one such a tunnel can be seen in the picture below. I walked up to the tunnel entrance, but did not go inside, suspecting boobytraps. From the military viewpoint, this airfield was tactically very well located with the high mountain range on its west which shields the airfield from the visibility of NATO radars based across the Adriatic Sea in Italy. Early in 1992, as the Yugoslav military was pulling out of Bihac, they cratered the runways to an unusable state.

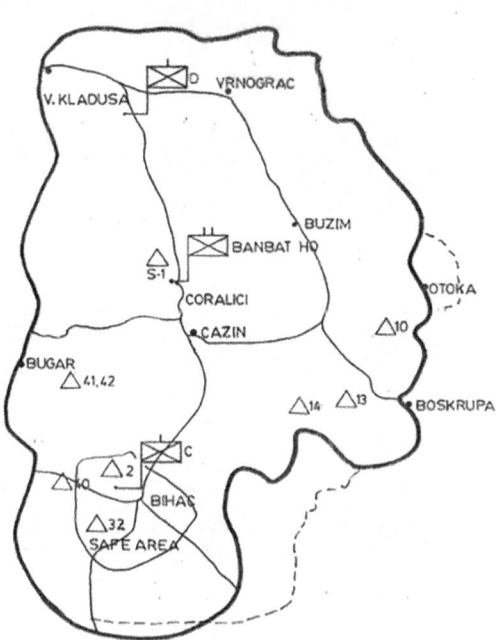

Bangladesh Battalion HQ in Bihac. A company at Coralici, a company with Logistic Base at Veika Kladusa and a company inside Bihac City Safe Area. The triangles show the location of the OP— Observation Post (Source: Author's own wartime records)

Former Yugoslav Air Force aircraft parking pens tunneled inside
Bihac mountains, Aug. 1995 (Source: Author's own records)

Author inside the former Yugoslav Air Force aircraft parking pens in
Bihac mountains, Aug. 1995 (Source: Author's own records)

.

Author visiting a Croat village south of Bihac City, 1995. Houses and Catholic church burnt by Serb forces (Source: Author's own records)

Bosnian Army 5 Corps soldiers picking up war booty after recapture of Bosanko Krupa in Bihac, Oct. 1995 (Source: Author's own records)

Bosnian Army Clandestine Air Supply Missions

For the Bosnian Army 5 Corps, supply of ammunition was a survival need. To meet the wartime demand, the Bosnian Army successfully converted a few local factories to help produce some desperately needed ammunition. The Bosnian Army in Bihac also had clandestine supply arrangements to meet their other warfighting needs. Unmarked MI-8 Helicopter flew twice or thrice a month at night into Bihac, dodging the Serb air defense to deliver the much-needed cargo. Single helicopter loads were small, may be equivalent to a truck load, but provided the critical warfighting needs. Sometime in Dec 1994, it was snowing heavily around 10 pm at night; I came out of my living container on hearing the sound of an incoming Bosnian night supply helicopter, an UNMO who was a helicopter pilot himself standing close to me was surprised to see an MI-8 helicopter flying in such adverse weather. He commented, *"This pilot must be flying on vodka in this atrocious weather with such heavy snowflakes pouring down."* Estimating the distance from the sound of the Helicopter, it appeared to have landed not too far from the UN camp. In a country which was completely dark at night due to the absence of electricity, the UN camps with bright flood lights probably served as a light beacon for night time navigation.

These night infiltration flights having to dodge the Serb air defenses were certainly very risky and dangerous. At times these flights also carried few passengers, and it was one such flight in May 1995 on which Bosnia's wartime Foreign Minister, along with a few others, came on a secret visit to Bihac. A few days later, on their return flight, the helicopter was shot down, killing all onboard, including the Bosnia Foreign Minister. I learnt that these nighttime flights originated from unknown locations in

Croatia, and the aircrew were paid mercenaries from some East European country. Over my long fourteen months stay in Bihac, the most daring and astonishing night infiltration flight was of a transport aircraft landing on a semi-prepared gravel strip at about the same location where the helicopters used to land. From the engine's reverse thrust sound, the aircraft sounded like the Russian transport aircraft AN-32 or AN-26. I have flown on both types of these aircraft and therefore my assessment. The cargo was delivered with the engines running, and the aircraft took off again after a short interval for the return flight.

UK Special Operations—SAS Teams

In Bosnia Herzegovina, UK was operating a few independent SAS (British Special Force) squadrons. These independent SAS squadrons were the national assets of the UK Ministry of Defense and not part of the UN Force. Unlike the peacekeepers, they did not wear UN blue headgear and drove around in UK military vehicles, not the white UN vehicles. However, they enjoyed logistical support from UN Camps, but operated independently. These small units were like eyes and ears for the UK Ministry of Defense to get first-hand reports directly from the field. Locally they appeared to be under the control of the UK General in command of the UNPROFOR, Major General Rose in 1994 and later Major General Smith in 1995.

One such SAS team with one Captain, three SAS soldiers and two Land Rovers, were stationed in my camp soon after the surge in fighting in the area. We provided them with logistical support like accommodations and fuel. This team would speak to us only if they needed something. I noticed they were well equipped, carrying satellite communication equipment,

Rockwell military-grade GPS and personal weapons. They patrolled the entire battle areas and probably had contact with the warring parties on both sides, liaising possibly through HQ UNPROFOR at Sarajevo. I assume the SAS team had full air support overhead at all times, or else they could not have operated independently in such small numbers in remote locations away from the UN peacekeepers.

UNPROFOR Established New Command HQ in Bihac

Further, with the surge in fighting in Bihac, HQ UNPROFOR Sarajevo established its direct monitoring in the area by establishing a new Command HQ named Bihac Command (BH Command). Bangladesh Battalion was placed under the new BH Command HQ. I remember the Deputy Force Commander at UNPROFOR HQ at Sarajevo, a Brigadier from Belgium, called me on the phone at night around 11 pm to inform me of the decision about the new Command HQ. We had some heated arguments as I questioned why was such a command not created in the past when the French Battalion was present and whether they were treating us differently for being from a non-NATO country. The Deputy Force Commander stated they were reinforcing the UN presence in the area given the worsening war situation. I told him that in a military sense, reinforcement would imply sending additional combat-capable units to enhance the deterrence value. The insertion of additional combat units would also justify the creation of a new Command HQ or Sector HQ. Obviously, no answer came from the other end. Colonel Limue, a Canadian colonel who was working on operational staff at UNPROFOR HQ in Sarajevo, was re-assigned as the new BH Commander. In my judgement, this new arrangement would not make sense to any military professional, a Command HQ to command a single battalion and a few UNMOs. Clearly, the arrangement was to strengthen the watch over the Bangladesh Battalion in Bihac. The new BH Commander understood that I did not accept the change well, so he kept his distance and came to see me after one week. I told him he could come and see me if there was any issue but should not expect me at his meetings. My staff officer attended the daily meetings and came back to brief me.

We used to send daily SITREPS *(Situation Reports)* twice a day to HQ UNPROFOR as per UN specified format. The SITREPS included details of ceasefire violations, incidents of firing, shelling, and civilian casualties. Since we were situated on the ground, we were the extended eyes and ears to report from the battle front line. HQ UNPROFOR would normally compile such reports from all sectors and send them to UN New York with their comments added. After a few weeks, I noticed that our reports, especially the part pertaining to ceasefire violations by the Serbs were blanked out in the final report going to UN New York. In protest, we started sending blank reports which caused HQ UNPROFOR staffs to query me. I asked them why they were leaving out our reports on Bihac to which they kept quiet. This indicated that either HQ UNPROFOR did not trust the reports from the Bangladesh Battalion or did not want the content on 'Safe Area' violation by the Serbs to be reported. May be somebody in the HQ did not like to see the Serbs as bad guys, may be this was part of the policy for some of the European nations at the time. Ongoing controversies and loose talks on why the Bangladesh Battalion was deployed in Bihac reached my ears. Therefore, I sent the report to HQ

UNPROFOR to which they never replied. They did not even call me to discuss my report.

Fikret Abdic, Rebel Faction Fights Bosnian Government Forces

In September 1993, Fikret Abdic, the Muslim rebel faction leader, declared the northwestern part of the Bihac enclave as an autonomous region.

Author (far left) meeting with Fikret Abdic after his forces were in temporary control of NW Bihac, 1995 (Source: Author's own records)

Fikret Abdic's business group called 'Agrokomerc' had many well-established agro-processing factories in the northern part of Bihac. These factories employed hundreds of locals, and therefore Fikret Abdic enjoyed the support of the locals in the northern part of Bihac. Being separated from central Bosnia, he wanted to continue his agro-business by exporting his processed agro-products to neighboring Serb-held areas and beyond to Croatia.

Doing business or keeping any relationship with the Serbs was not acceptable to the Bosnian central government in Sarajevo, which resulted in the split.

THE DECLARATION
to pronounce Republic of Western Bosnia

1. Autonomous Province Western Bosnia is hereby proclaimed Republic Western Bosnia within the borders of internationally recognize Bosnia Herzegovina. Mutual bordering will be executed by an agreement with the other members of Bosnia Herzegovina.

2. Republic Western Bosnia will be a neutral member of Bosnia Herzegovina.

3. Internal organisation of Republic Western Bosnia will be regulated according to its specific features and interests of its citizens.
   Human rights and liberties will be regulated by rules and practically excercised according to the valid international documents.

4. Finalisation of political and constitutional crisis of Bosnia Herzegovina will be excuted by a political agreement of the socio-political communities that appeared in the time period 1992-1995 with the help of international community.

5. Permanent end to the war and animosities within Bosnia Herzegovina should be achieved by an Agreement for permanent peace with the guarantees of international community.

6. There are existing peace documents signed with the encirclement as follws: the Common statement signed 21 OCT 1993 with the President of Croatian Republic Herceg Bosna, the Declaration signed on 22 OCT 1993 with the President of Republic Srpska and the Statement signed 07 NOV 1993 by the Prime-ministers of Croatian Republic Herceg Bosna, Republic Srpska and Autonomous Province Western Bosnia.

7. Ruling Bosnia Herzegovina in the interim should be regulated by a Security Council Resolution or a written Agreement of the authorized representatives of the socio-political communities.

8. In the area of Republic Western Bosnia the rules of Bosnia Herzegovina and rules of Autonomous Province Western Bosnia will be applied, if they are not contrary to this Declaration.

9. Bodies of Republic Western Bosnia will be appointed according to the rules of Autonomous Province Western Bosnia within 30 days from the day of proclamation of this Declaration.

10. The temporary Constitution of Republic Western Bosnia will be made by the Republic bodies within 60 days from the day of proclamation of this Declaration.

11. The president of Republic Western Bosnia and the Republic bodies will ensure the fulfillment of this Declaration.

12. The Declaration will be published in the Official Register of Republic Western Bosnia.

President of
Republic Western Bosnia
Fikret Abdic

The Bosnian central government in Sarajevo ordered the Bosnian Army 5 Corps to crush the rebellion by Fikret Abdic and his supporters. Bosnian government forces successfully defeated Fikret Abdic's forces after about one year in August 1994. Fikret Abdic, with around 30,000 followers, fled the pocket and became refugees camping at Turang and Batnoga across the border in the RSK area *(Self-declared Republic Serb Krajina—Southern Croatia).*

BanBat Velika Kladusha Camp sandwiched between Krajina Serb forces on the photo backside and Bosnian Government 5 Corps on the hills seen on the far side. Serbs target the lead BanBat APC as it was entering the camp, 1994 (Source: Author's own records)

In the refugee camps in Krajina Serb territory, Fikret Abdic's forces became a good tool in the hands of Serbs to be trained and rearmed to make them fight the Bosnian Army 5 Corps, an objective that served the bigger interest of the Serbs. Together with the RSK Serbs, Fikret Abdic forces jointly attacked Bihac for the second time in December 1994 and managed to re-capture the northern area. Serb forces led the attack with the extensive support of artillery guns firing from across the border. During the final stages of the battle, the

Bangladesh Battalion camp at Velika Kladusha was sandwiched between Abdic Serb forces and the Bosnian Army 5 Corps. Serb forces had positions as close as 300 meters on hills on the west. Bosnian Army 5 Corps soldiers had positions on the high grounds in front of the UN camp on the hills on the far side in the picture. During a pause in the battle between the Serbs and the Bosnian Army 5 Corps on December 12th, 1994, two of our APCs were returning to our Velika Kladusha Camp after a patrol mission. The Serbs targeted our APCs in front of our camp entry gate. The lead APC was hit twice by Serb-fired ATGM (*antitank guided missile*). The first ATGM hit the gun turret, which was blown away, killing a soldier Mohd Ismail. This APC caught fire, and as other soldiers rushed to evacuate the injured, Serbs fired a second ATGM, which hit the APC front left and injured a few more soldiers inside the APC.

Closeup view of the BanBat APC targeted by Serb fired ATGM as it was entering the VK Camp, 1994 (Source: Author's own records)

The Serb attack on UN soldiers was immediately reported and escalated to the highest level at UNPROFOR with the request for intervention. I was told that NATO aircraft were overhead, and I could request a NATO air strike on the suspected Serb location. Such a request would require me to pass on the GPS coordinates of the suspected Serb position. My people at the site did not have any GPS on hand; further, getting the correct coordinates in evening darkness was difficult. Airstrike on a wrong target could have other implications. At the time, my priority was evacuating the wounded immediately to save lives. So instead of asking for an air strike, I asked for an immediate ceasefire from the warring parties to enable me to evacuate the wounded to Zagreb. Evacuation required first passing through the battle lines and after that, a night journey of a few hours, passing through territory held by Croatian Serbs. It was winter and late in the evening; however, with UNPROFOR's quick intervention, the Serbs and Bosnian Army 5 Corps agreed to cease fire, allowing me to evacuate the injured immediately. In the middle of the battlefield with trigger-happy soldiers, we could only use armored ambulances. Luckily we had a few US M113 armored APC ambulances which were provided by UN and these ambulances came to good use for evacuation through the contested battle lines.

After picking up the casualties, the ambulance drove through the Serb and Bosnian Army 5 Corps front-line fighting units, which had been exchanging fire an hour ago. Fortunately, the evacuation went without any further incident, and the four injured soldiers reached the US Mobile Army Surgical Hospital (MASH) hospital in Zagreb, Croatia, some 100 km away. The decision to evacuate quickly saved the lives of a few of my soldiers. In hindsight, I can say that calling for the air strike at that particular moment would have been a bad decision because it would have prevented the evacuation of the wounded soldiers and could have escalated the fighting.

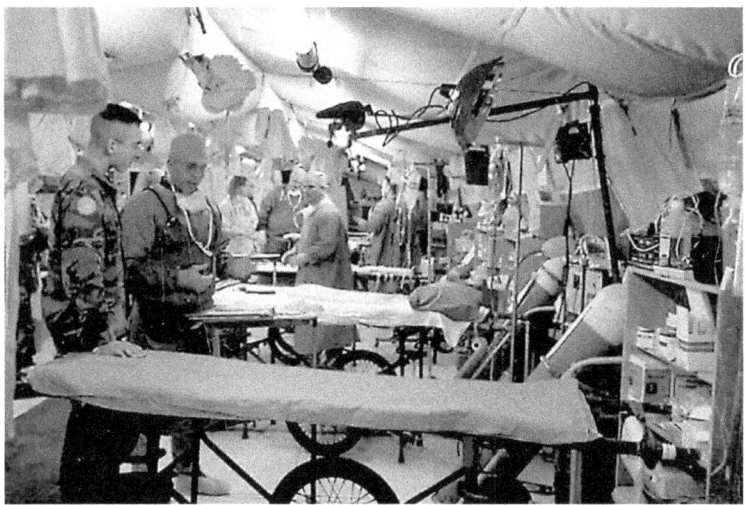

After the Serb missile hit on BanBat APC, injured soldiers were evacuated to US Army Field Hospital, MASH (Source: Author's own wartime records)

Additionally, for a successful air strike, we were required to positively identify the target, which was difficult. We only saw the missile hit on the APC, but did not see the missile launch. While we knew the suspected hill from where the missile had come from, we could not be sure of its position. It was night; darkness had fallen; our soldiers on the ground did not have night vision binoculars or any other means to laser designate the target. Serb positions were within a few hundred yards, and any error on air strike could have caused collateral damage to our camp and the peacekeepers. Ideally to take an air strike in close proximity of the soldiers we should have had an 'Forward Air Controller Team' on site with required communications with the NATO pilots. So any air strike with no controller on ground, no GPS, no night vision goggles and no laser designator to illuminate the target would have ended in fiasco and likely collateral damage to UN camp which was withing few hundred yards.

Bosnian Army 5 Corps' attack on Abdic Forces, Aug. 1995
(Source: Author's own wartime records)

CAPTURED EQUIPMENTS

1.   03 X 122 MM GUN/HOW.

2.   15 X 105/76 MM GUN.

3.   07 X 120 MM MORTARS.

4.   02 X 76 MM ZIS GUN.

5.   01 X 85 MM M-44 TANK.

6.   02 X T-55 TANKS.

7.   03 X 130 MM GUN.

8.   03 X 20 MM M 75 AA GUN.

9.   02 X 30 MM M 53 SP AA GUN.

Bosnian Army 5 Corps capture RSK weapons in Fikret
Abdic Battle for Velika Kladusha, Aug. 1995 (Source:
Author's own wartime records)

UN Sector North, South, West, and East. Self-declared autonomous areas under control of Croatian Serbs until August 1995 (Source: Author's own wartime records)

Krajina Serb refugees going to Serb held areas in Bosnia, Herzegovina, 1995 (Source: Author's own records)

Krajina Serbs discard weapons and head for Serb held areas in Bosnia, Herzegovina, Aug. 1995 (Source: Author's own records)

Krajina Serbs discard ammunition and head for Serb held areas in Bosnia, Herzegovina, Aug. 1995 (Source: Author's own records)

COMMAND OF THE 1st NOVI GRAD   MILITARY SECRET
INFANTRY BRIGADE   Strictly confidential
strictly conf.No.03/278-1459
Date: 2 November 1994

Criminal charge for the
betrayal, desertion and
non-fulfillment of orders,   2nd battalion

On the basis of the order of 1st Krajina Corps strictly conf.
No.1/1-219/1 of 1 November 1994, in accordance with the declared
state in war in the area of the 2nd Krajina Corps responsibility
and the elevation of fighting spirit in the RSA units on the
highest level, with consolidation of the RS defence and breaking
of the enemy offensive as the goals, and on that basis the
elevation of responsibility, military discipline and professional
relationships on the higher level

I ORDER:

1. The measures of a criminal charge and procedures against all
senior officers acting from the duty of the battalion commander
to the duty of the platoon commander who are responsible for the
self-willed abandonment of the defence positions, loosing the
territory without a fight against the enemy, casualties and
material damage, should be undertaken.

2. The abandonment of positions, desertion and all other kinds
of behavior that is not according to the military orders and
without the approval of the HQ RSA Commander and Corps Commander
regardless of the possible casualties and consequences, is most
strictly forgotten.

3. The senior officers of all levels must strengthen the morale
of soldiers and inspire the confidence and safety to population
into the Serb Republic Army by their firmness, stubbornness and
professional relationship in the accomplishing of ordered tasks
as well as by their behavior during the fight.

4. For the realization of this order the commanders of battalions
and independent units are personally responsible to me.

Date: 3 November 1994   COMMANDER:
    received by: Lieutenant-Colonel
1st unit: Bukvic S. (signed)   Ranko Dabic
2nd unit: Vukojevic Z. (signed)   (signed and sealed)
3rd unit: Bujevic B. ------
MM unit: Ciric V.(signed)

Translated copy of Serb's first Novi Brigade orders warning Serb soldiers not to abandon or desert positions, 1995 (Source: Author's own records)

Fikret Abdic's dream of an autonomous republic of Western Bosnia was short-lived. During the few months when Abdic was in control of the area, I had the opportunity to meet with Abdic and his Chief of Staff. During the meeting, I had to bear with the long political talk of Abdic, after which I was given a copy of the declaration document for 'Autonomous State of North Western Bosnia. It was difficult for Abdic's forces to fight the Bosnian Army 5 Corps on their own, so they relied on the active involvement of the Croatian Serb Forces—RSK. The motivation for the Serb forces to get involved was to defeat the Bosnian Army 5 Corps in Bihac, which had become a formidable resistance force. Reportedly Abdic was paying cash to the Croatian Serbs for the support he was getting. On August 7th, 1995, Bosnian Army 5 Corps (*Bosnian government forces*) attacked and defeated the rebel Abdic Forces and their allies, the Croatian Serb forces and recaptured the entire northern area of Bihac. As a result, Abdic forces withdrew across the border into Croatia to become refugees for the second time and remained so till the end of the Bosnian war. The Commander of the Bosnian Army 5 Corps, General Dudakovic, arrived at Velika Kladusa in the afternoon to celebrate the victory. After the war, Abdic was arrested and tried in Croatia for war crimes and convicted with a jail sentence.

Bihac On the Brink of a Genocide

Bihac being encircled on all sides by the Serb forces was dependent on UNHCR convoys to bring in the needed food and medical supplies but were not getting permission from the Serbs to travel to Bihac. Bosnian Serbs, under their Army Commander General Miladic, attempted twice to capture Bihac. Miladic, first attempted to capture Bihac around mid-Aug 1994, jointly with their allies RSK—Serb Army from Krajina (*Southern Croatia*). They attacked from the east around the Buzim area. Serbs could capture significant ground initially, but later, the Bosnian Army 5 Corps counterattacked with two brigades and recaptured the lost ground forcing the Serb Forces to pull back. This battle saw fierce fighting, with both sides suffering heavy casualties. There were also reports that General Mladic's personal dairy fell into the hands of Bosnian Muslims 5 Corp, which, if true, indicates that General Mladic came close to being captured or killed. The Serbs withdrew from the area after losing the battle and reportedly taking heavy casualties.

A second attempt by the Serbs to capture Bihac took place around mid-Nov 94 when the Serbs attacked Bihac city from the south. This time the Serbs gained a lot of ground, including encroachment into the UN-designated Bihac Safe Area. As the Serbs advanced, they burned villages, demolishing a mosque at Sokolac and a catholic church in the Croat village of Zaveleje. Serbs also captured some vital high ground overlooking Bihac city; from these positions, Serbs kept targeting the population by sniping and shelling. Every day, scores of injured civilians, including children, were admitted to Bihac Hospital for treatment of injuries. The Serbs also captured the Klokot water reservoir cutting off the water supply to the city, which caused unbearable suffering to the population. The Serbs had reached the outskirts of Bihac city and were poised to attack and capture it. HQ of the Bosnian Army 5 Corps and other offices of the Bosnian government were all located inside Bihac city. Attacking the city would have certainly resulted in high civilian casualties, as people were cramped in the city coming from surrounding villages. The Serbs gave a deadline to the Bosnian Army 5th Corps to surrender, which was rejected by the Bosnians. The city started building street blockades and getting ready for street fighting. The Bosnian Army 5th Corps and other government offices started relocating and destroying important documents to prevent capture by the Serbs. The Bangladesh battalion had a camp with about a hundred soldiers inside Bihac city at the Bihac School premises. Everyone expected Bihac city to fall anytime to Serb hands, and the worst fear was that the tragedy of Sebrenecia could be repeated in Bihac. My assessment was if the Serbs did capture the city, they would be hostile towards the UN Peacekeepers and, in all likelihood, would cut off the UN supply routes and restrict movement. So, to prepare for such an eventuality, I sent a truckload of food items, compo ration packs, and fuel so the camp could hold on and survive the crisis. Fortunately, this attack on the city did not come as the Bosnian Army 5th Corps launched its own counter-offensive and pushed back the Serbs far to the south. Copied below is the operational sketch, which I had drawn and sent with my situation report at the time to UNPROFOR reporting on the Serb encroachment into the UN declared Bihac Safe Area.

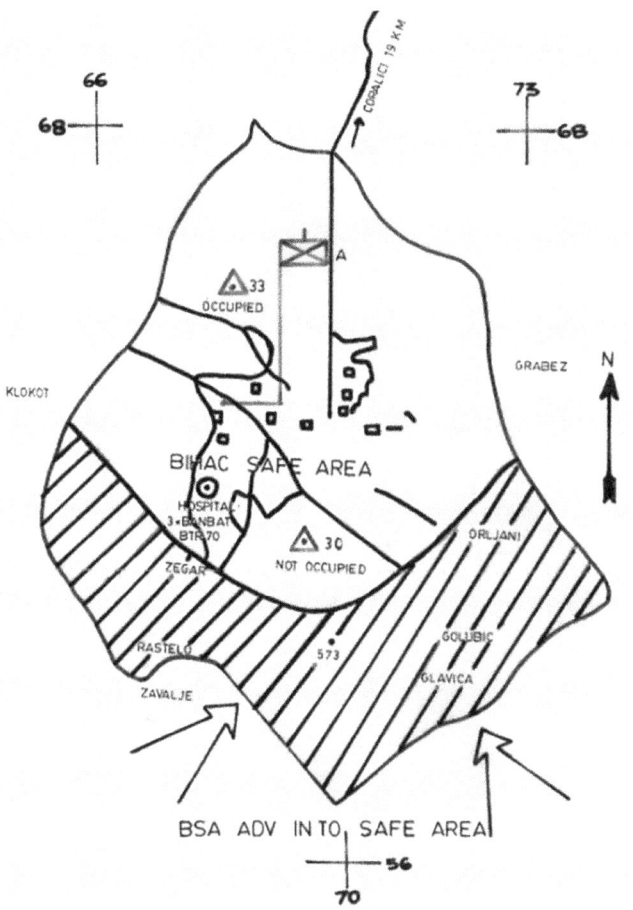

Shaded area shows southern part of the city captured by Bosnian Serb Army (BSA). BSA gave an ultimatum for Bosnian government forces, 5 Corps, to surrender. City came close to being overrun by BSA which was averted after 5 Corps counter-attacked and pushed back the Serbs, Feb, 1995

This is an overlay used with military maps, and therefore, one can see the map alignment cross at three points. However, even without the map, it is useful, as the shaded areas show the extent of Serb incursion into the Safe Area and the proximity of Bihac hospital. Bangladesh

Battalion APCs were positioned on the outer edge of the hospital on 24/7 watch to safeguard the hospital from falling into Serb hands, an operational task which we successfully completed.

In the sketch, the outer line roughly depicts the city limits. The shaded area shows the extent of Serb incursion into the UN-declared safe area in December 1994. Note the location of Bihac Hospital on the edge of the shaded area where three Bangladesh BTR-70 APCs were positioned to protect the hospital. The Blue flag shows the location of the Alfa Company Bangladesh Battalion, and the other Blue Triangle is the Observation Post which was manned 24/7. In December 1994, Bihac city came close to being overrun by the Serbs; fortunately for the people of Bihac, another mass genocide like Srebrenica was averted due to a counterattack by the Bosnian Army 5 Corps.

Author (right) briefing UNPROFOR Commander, Gen. Rupert Smith, 1995 (Source: Author's own wartime records)

CO BanBat (author) visiting troop locations in the confrontation line, 1995 (Source: Author's own wartime records)

Cazin residential area near Corallici bombed by Bosnian Serbs (Source: Author's own wartime records)

Bihac enclave had only one hospital for its nearly 180,000 wartime population. Almost every day, women and children wounded due to the Serb shelling kept pouring into the hospital, thus straining the hospital resources to unbearable limits.

Resilient Bangladesh Soldiers Protect Bihac Hospital

BanBat APC deployed for protection of Bihac Hospital facing the Bosnian Serb forces after they encroached Bihac Safe Area threatening to overrun the city (Source: Author's own wartime records)

Deployment of APCs at Bihac hospital in the middle of the Yugoslavian winter, on 24/7 watch, did present its challenges. We were in the middle of the infamous 10-week Serb blockade period and running very low on fuel reserves. In the winter, even on static duty like at the hospital, the APCs with crews staying inside would frequently run their engines for heat and to charge the notoriously low-quality BTR-70 batteries, especially at night. However, due to fuel rationing, our APC crews were instructed to limit the engine static run time to save fuel. APC is like a metal can and to be inside the APC in cold weather without heat is intolerable. Due to fuel shortage, Bangladesh APC crews were subjected to untold suffering.

They often complained of cold-induced pain shooting up their calf muscles. The picture shows a soldier half exposed on the APC, exposing himself to freezing temperatures. The purpose of keeping the soldiers visible was to send a message to the Serbs that firing at the hospital would be seen as targeting the UN, which they knew was the trigger for calling NATO air attacks. For soldiers from the Asian tropical region who had never seen snow in their life, this was asking for a display of extreme resilience, human tolerance, and military discipline. On the battle front, apart from the fighting on the city outskirts, Serbs kept bombarding the civilian populated areas with long-range artillery guns and probably also surface-to-surface missiles.

UN Camps and the Observation Posts, OPs, were subjected to frequent shooting and shelling, and many of us in vehicles or APCs have experienced being shot at. I can recall on one occasion sometime in December 1994, when I was crossing the confrontation line between Abdic and Serb Forces and the Bosnian Army 5 Corps, my Land Rover took six hits from small arms. The Serb threat to Bihac city ended with the successful counterattack by the Bosnian Army 5 Corp on January 14th, 1995. Bosnian Army 5 Corps re-captured Klokot water reservoir and areas further south, including Ripac, where the HQ of Serb 2 Krajina Corp, Croatian Serbs, was located. Bosnian Army 5 Corps success in pushing the Serbs far south freed Bihac city from direct targeting and shelling. Bosnian Army 5 Corp's counterattack from Bihac resulted in a significant loss of territory to the Serbs.

Bihac Command Report on Serb Forces
Positioned close to Bangladesh Battalion VK Camp
Incidents of Serb Firing on Camp
(Source : Author's own war time records 1995)

1. VISITED TODAY BANBAT VELIKA KLADUSA CAMP TO ASSESS MILITARY SITUATION AN[D] THREAT TO BANBAT SOLDIERS. IN BRIEF, THE FIGHTING IS LITERALLY OUTSIDE THE FENC[E] AND THE NUMBER OF INCIDENTS INSIDE THE CAMP ENCLOSURE IS INCREASING WHICH MAKE[S] ME CONCLUDE THAT UNPROFOR NEEDS TO BE PROACTIVE WITH RSK (KNIN AND BANJNA CORPS[)] TO GET THEIR FORCES TO REDEPLOY INTO AREAS THAT ARE NONE THREATENING TO UNPROFO[R] AND TO ESTABLISH REGULAR LIAISON BETWEEN UNMOS BIHAC AREA AND LOCAL RSK GROUN[D] COMMANDER TO RESOLVE ISSUES IMMEDIATELY.

2. FIGHTING HAS CLOSED UP TO THE BANBAT CAMP WITH RSK FORCES THAT ARE ON THE HIGH GROUND WEST OF THE CAMP AND IN DEFENSIVE POSITIONS (COUNTED SIX) ON IT[S] FORWARD SLOPE (300M-600M FROM THE CAMP FENCE). THE FOLLOWING SUMMARIZES THE FIRING IMPACTS INSIDE THE CAMP (DELIBERATE OR OTHERWISE):

DATE	TYPE OF ATK	DETAILS
19 NOV 94	MORTAR	ONE SHELL THROUGH WAREHOUSE IN LOGBASE.
20 NOV 94	TK SHELL	SHELL HIT CONTAINER INSIDE BANBAT LOG BASE.
20 NOV 94	SAF	ELECTRICAL REPAIRMAN FIRED AT WHEN TRYING TO REPAIR DAMAGE FROM TK RDS
26 NOV 94	SAF	SNIPER FIRE IN BANBAT LOGBASE. HIT A SLEEPING CONTAINER WITH TWO BULLETS.
27 NOV 94	ARTY	2 X SHELLS DROPPED INSIDE BANBAT PERIMETER AT THE LOG BASE
05 DEC 94	SAF	STRAY BULLET INSIDE BANBAT LOG BASE
05 DEC 94	SAF	ASSESSED SNIPER FIRE TOWARDS TWO BANBAT SOLDIERS INSIDE THE LOGBASE PERIMETER.
07 DEC 94	SAF	THREE STRAY BULLETS. ONE HIT A COOKHOUSE, ANOTHER A STORE CONTAINER AND THE THIRD BROKE THE WINDOW OF A 2 TON LORRY
08 DEC 94	SAF	BULLET HIT WATER TRAILER
08 DEC 94	ARTY	TWO SHELLS IMPACTED INSIDE THE LOG BASE

3. WE HAVE WORKED WITH SECTOR NORTH TO KEEP THE RSK LO AWARE OF OUR CONCERN FOR THE BANBAT CAMP. HOWEVER NOW I BELIEVE WE REQUIRE A HIGHER LEVEL O[F] INTERVENTION BEFORE THE FIGHTING INCREASES AS RSK HAVE NOT ACHIEVED THEI[R] OBJECTIVES AND WINTER WEATHER CONDITIONS WOULD GREATLY AFFECT THE ATTACKER[S] ABILITY TO SUCCEED, AND ALSO, THIS WOULD LEAVE THE ABDIC REFUGEES IN THE KRAJIN[A] CAMPS IN DIFFICULT LIVING CONDITIONS. THE FIGHTING COULD THEREFORE INCREAS[E] CONSIDERABLY DEPENDING ON THE INVESTMENT BY RSK FORCES WHO WILL LIKELY AT THI[S] POINT TURN TO TACTICS SEEN IN THE SOUTH OF BIHAC, IE DESTROY THE DEFENDED BUIL[T] UP AREA IF NEEDED TO SUCCEED. (PLSE NOTE THAT THERE HAVE BEEN CLEAR TARGETING O[F] CIVILIANS IN VELIKA KLADUSA AND SURROUNDINGS, AS WITNESSED PERSONALLY TODAY B[Y] COMD BAC JUST OUTSIDE BANBAT CAMP ENTRANCE.)

4. THE THINNING OUT OF VELIKA KLADUSA STARTED TODAY WHICH WILL CONTINUE LEAVIN[G] LESS BANBAT SOLDIERS AT RISK. ALSO HAVE TASKED TACP TO CONTINUE TO LOOK FOR GOO[D] OPS TO OBSERVE GROUND TO THE WEST OF BANBAT CAMP. (A TANK AND AA GUN HAVE BEE[N] OBSERVED FIRING FROM THESE LOCATIONS.)

Report initiated by Bihac Area Commander, page 1, Mar. 1995
(Source: Author's own wartime records)

WOULD APPRECIATE THE FOLLOWING INTERVENTION TO BE MADE BY NEGOTIATORS WITH RSK KNIN (UNPROFOR HQ) AND RSK BANJA CORPS (SECTOR NORTH) TO MAKE THE POINT THAT:

A)   THE BANBAT CAMP AT VELIKA KLADUSA HAS BEEN THE SUBJECT OF TOO MANY FIRING INCIDENTS AND THAT THIS MUST CEASE.

B)   DEFENSIVE POSITIONS THAT HAVE BEEN PREPARED ON THE FORWARD SLOPE WEST OF THE CAMP MUST BE REMOVED TO AN AREA WHICH DOES NOT PUT THE CAMP AT RISK OF CROSSFIRE OR DIRECT ATTACK.

C)   A HOT LINE BE ESTABLISHED BETWEEN SECTOR NORTH AND BANJNA CORPS TO ENSURE THE QUICK RESPONSE TO INCIDENTS.

D)   THE LOCAL RSK FIELD COMMANDER OR HIS REPRESENTATIVE BE AUTHORIZED TO MEET WITH UNMOS BIHAC AREA TO ESTABLISH A MEANS OF CONTACT, IE AT A POINT TO THE WEST OF THE BANBAT CAMP, FOR INSTANCE.

E)   PLANNING FOR THE USE OF CAS IN SUPPORT OF THE BANBAT CAMP BE INITIATED WITH NATO IN ORDER THAT THE DIFFICULTIES OF OPERATING NEAR OR IN CROATIA BE EXAMINED AND SOLVED.

F)   RSK SHOULD BE ADVISED OF THE CONCERNS BEING EXPRESSED HERE AND THAT UNPROFOR IS REVIEWING ITS CAS PROCEDURES WITH NATO CONCERNING THE USE OF AIR IN THE AREA.

6.   YOUR ASSISTANCE IS REQUESTED.

**Report initiated by Bihac Area Commander, page 2, Mar. 1995
(Source: Author's own wartime records)**

The Bosnian Army from Bihac captured a huge land area and a lot of equipment from the retreating Serb forces, including tanks and artillery guns. Serbs again started blaming the Bangladesh Battalion for their battlefield losses, even accusing Bangladesh soldiers of helping the Bosnian Army with grenades and ammunition. The only evidence Serbs produced to the media were a few hand grenades with Arabic writings, and they assumed the Muslim connection. To counter Serb propaganda, I did my media campaign to show that our small arms, ammo, and grenades were all made in Bangladesh with English inscriptions. Back in Bangladesh, families remained worried about our safety because of the worsening battle situation in Bosnia being reported on television. I routinely sent my reports to Military Operations Directorate in

Bangladesh; one such report is copied below. At the bottom of the last page, note the author's handwritten remarks at the end. UNPF Zagreb, took an unusual step by barring all telephone calls going out of the Bangladesh Battalion in Bihac for several days in November 1994. This was possibly done to stop information from going out of Bihac. Such drastic action could only be taken on instructions from HQ UNPF as we were on UN communications backbone.

Serb Blockade of Bihac—All Troops Are Not Equal

Serb reaction to the Bosnian Army 5 Corp attack was to punish the UN Peacekeepers in Bihac by imposing a complete blockage to the entry of UN supply convoys to Bihac. Even the UN support helicopters were not given flight clearance and were barred from flying into Bihac. Between October 1994—December 1994, Bangladesh Battalion experienced some ten weeks of blockade on all supply convoys. The battalion had to rely on its meagre stock of fuel and food. I had to impose severe rationing, especially on hours generators could be run for heating and lighting. Lack of heating during the winter caused immense suffering to soldiers of the Bangladesh Battalion. Fuel shortages also restricted operational activities as APC operations consumed a lot of fuel. At this time, I had some unpleasant discussions with the UNPROFOR staff for their failure to get in our logistical support convoy. I always blamed the UNPROFOR leadership for total surrender to the Serb's rogue behavior. I loudly voiced my views questioning: had there been a battalion from any of the European countries facing the blockade would UNPROFOR and NATO sit back and watch as they did in our situation. Maybe we were being treated differently simply because we were from another part of the world. My interviews on the subject were published in the USA in early 1995 in LA Times and Philadelphia newspapers. I also spoke at the time on Sky TV UK and on my interview on Peter Jenning's talk show aired in the USA on ABC TV on March 25th, 1995. UN New York was not happy with me for airing my views on the media and mentioning the UN's failures. As the commanding officer of approximately twelve hundred soldiers, I was more concerned about their suffering than the disapproval of the UN regarding my media interviews.

Fortunately for us, the French battalion was oversupplied, and they had left behind a few containers of French military compo ration packs. The French had also left behind a warehouse full of other tinned food items, including tons of water packs which came to good use to help us survive the ten weeks of the blockade. My soldiers noticed that some of the French military compo food packs had pork content which caused some discontent among my soldiers for religious reasons. Under the severe food crisis, Bangladesh Battalion learnt to adapt to whatever was available. My instructions to the soldiers was to throw out the can with pork meat and eat the rest, and everybody after that followed this survival need instruction.

I learnt an important lesson in Bosnia: food habits can severely dampen men's morale, whether soldiers or officers. Denying a person his staple food can be depressing, can demoralize the person and can even weaken the person mentally. A few weeks into the blockade, I was told by the Regimental Sergeant Major that the soldiers were depressed and had no energy to work because they had not been given a proper meal for weeks. It took some time for me to make sense of it, only after I found my officers had the same complaint and explained that they were missing their rice meal. Staple food like rice was more important than whatever calories we gave on the compo meals. The Bangladesh Battalion had carried a small quantity of rice, dal (*lentils*) and a few other dry items from Bangladesh. Due to the blockade, I had instructed earlier that we must preserve the stock of rice and use compo meals. I realized the absence of rice meals was mentally breaking down the soldiers. I allowed rice, dal, and aloo bhorta *(tinned potato meshed with green chilli)* once a week on Sundays. With the first rice meal, both the officers and soldiers were smiling again. Smokers had a hard time too, buying a pack of 20 cigarettes for $20, which was very expensive at the time. People smoked in groups taking a puff and passing the cigarette to the next person sharing one cigarette and going in circles. Some smokers even resorted to smoking cigarettes they made with tea leaves in rolled paper.

File Copy - Situation Report sent to Army HQ Bangladesh 19 Nov 1994

CONFIDENTIAL

HQ BANBAT BIHAC - CORALICI
TEL: VSAT - 385 41 180011 EXT 2282 OR 2283
FAX: VSAT - 385 41 180011 ASK 2478 TRANSMIT
TEL/FAX: TELE DATA - 001 440 988 2809 LONE TONE 8964
(TELE DATA CALLS ARE ROUTED THROUGH USA)

TO: DMC   FAX: 88 02 883144
      AHQ(MO DTE)

REF: BANBAT/1005/6   19 Nov 94

MESSAGE

SUBJECT: SITUATION REPORT BANBAT AOR

Ref:
A. Sketch of AOR sent earlier vide FAX dt 12 Nov 94.

1.   Situation in the AOR has further deteriorated. Bosnian 5th
Corps has lost more ground. In the south they are almost back on
the old confrontation line from where they had started. They are
being pounded by Serb guns. Although they have some guns, but
little or no ammunition to fire back. North West of Bihac city is
under attack by RSK(Self proclaimed Republic of Serb Krajina) and
Abdic forces(see notes) and they have taken approx 2-3 kms of land.
However the main fighting in the last 48 hrs has been in the north
centered around the town of Velika Kladusa. Mr Abdic was elected as
MP from Velika Kladusa Opstina(dist). So his forces are interested
in taking that town first(political requirement). Aided by tanks
and RSK soldiers, Abdic soldiers have gained ground upto northern
edge of the town. They have also infiltrated a little south and
have cut the road between Bn HQ and Velika Kladusa(our logistic
base and B company location). I have restricted move of our patrols
in the area due to firing going on there.   .

2.   The most interesting event of the day was an air attk by 4
Serb aircrafts which took off from UDBINA(Ref A). Though media
reported only two aircrafts, but Capt Afzal on our OP 30 which was
on the line of attack clearly counted 4 aircrafts. They targeted HQ
5 Corps located inside Bihac town. One Cluster Bomb and one Napalm
Bomb was dropped. The cluster bomblets were spread all over the
place but most did not explode. The Napalm bomb exploded spreading
gelatine around but did not ignite, therefore the bombs did not do
any damage. We withdrew OP 30 soon after that attack.

      Comment. The aircrafts had attacked Bihac, a UN designated
      Safe Area, made two passes in a NO FLY ZONE and surprisingly

CONFIDENTIAL

CO BanBat Situation Report to AHQ, page 1 (Source: Author's own wartime records)

NATO AWACs over the Adriatic sea missed them. This is
unbelievable with the technology onboard the AWACS. This was
the second attack in this month. On both occasions the NATO
patrol aircrafts were to be seen only after the attacking
aircrafts had left.

3.   Notes on Mr Abdic. A member of old Yugoslav Communist Party,
he is a businessmen by profession. He use to run group of Agro-
commerce companies which had multi- million dollar business in
Europe. He was elected MP from the opstina of Velika Kladusa. In
Sept 93 he choose to break away from the central govt and declared
Bihac as the 'Autonomous Province of Western Bosnia(APWB)'. He
created his private army of about 6 brigades.(a brigade here can be
anything between 1000 to 3000 men) He bought his weapons from
Krajina Serbs. He probably joined the serbs because his business
market was in Europe and he needed a corridor through the Krajinas.
On orders from the Bosnian government, 5th Corps launched its
offensive in Oct 93 and finally defeated Mr Abdic forces in Aug 94.
Mr Abdic with about 30,000(some estimates show a much higher
figure, the higher the figure the more aid one can plead for.
Higher figures also indicate that much more support Mr Abdic has.)
supporters crossed into Krajina and were so long in Refugee camps.

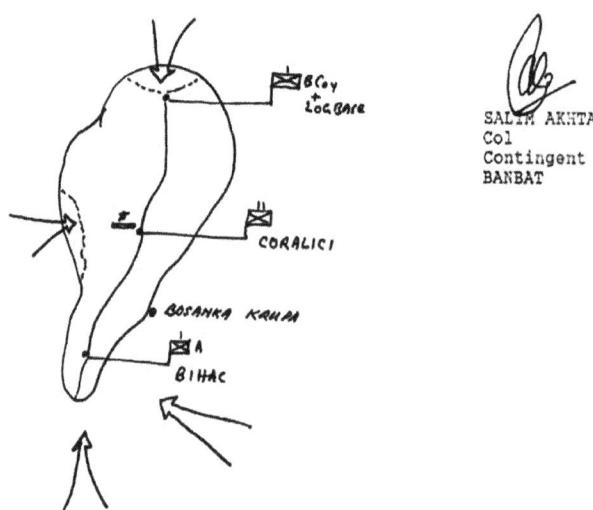

SALIM AKHTAR
Col
Contingent Comd
BANBAT

CoralIci airstrip

NOTE :   HQ ZAGEB is not putting cells through to
Bihac specially those coming from Bangladesh. Pl
info MOFA. Call us on Tele data nos.

**CO BanBat Situation Report to AHQ, page 2 (Source: Author's own
wartime records)**

August 1995 Joint Croat— Bosnian End of War Offensive

In August 1995, The Croatian Army and Bosnian Army, 5 Corps launched a joint attack targeting the Krajina Serbs in in Southern Croatia from multiple directions. This attack resulted in breaking the Serb encirclement of Bihac. Taking advantage of the Serb military's chaotic situation, the Bosnian Army 5 Corps started a fresh attack in the south, recapturing huge territory, as shown in the sketch below. The Red line indicates the confrontation line until August 1995, and the dotted GREEN & RED line shows the extent of the new area captured by Bosnian Army 5 Corps when the war ended. Do note that the Bosnian Army offensive stopped short of capturing the Serb stronghold city of Prijedor close to Banja Luka, both being large and politically important cities. The Bosnian Serb Air Force was also based in Banja Luka. General Dudakovic, Commander of the Bosnian Army 5 Corps, would have wanted to capture Prijedor, which I believe was his home town before the war. If Banja Luka had been lost, the Bosnian Serbs would have been reduced to an insignificant political partner in the emerging Federation of Bosnia Herzegovina.

Certainly, the European lead decision-makers did not want this to happen. They wanted the Serbs, Croats, and Bosniac Muslims in the new nation to be politically and geographically balanced to counter each other's clout. To the Europeans, the bigger concern was for Bosnia not to have an Islamic character. To stop the Bosnian Army from entering the two cities of Prijedor and Banja Luka, UNPROFOR warned the Bosnian Army that NATO would bomb them if they did not stop. So, the line was drawn and settled as the internal boundary between the Bosnian Muslims and the Bosnian Serbs in the new state of Bosnia Herzegovina. Further, in the south Croat forces attacked the Serbs from the west and linked up with the Bosnian Croat Army in Central Bosnia. This provided the Bosnian Croats a vital corridor to Croatia with proper and direct access to the Adriatic sea.

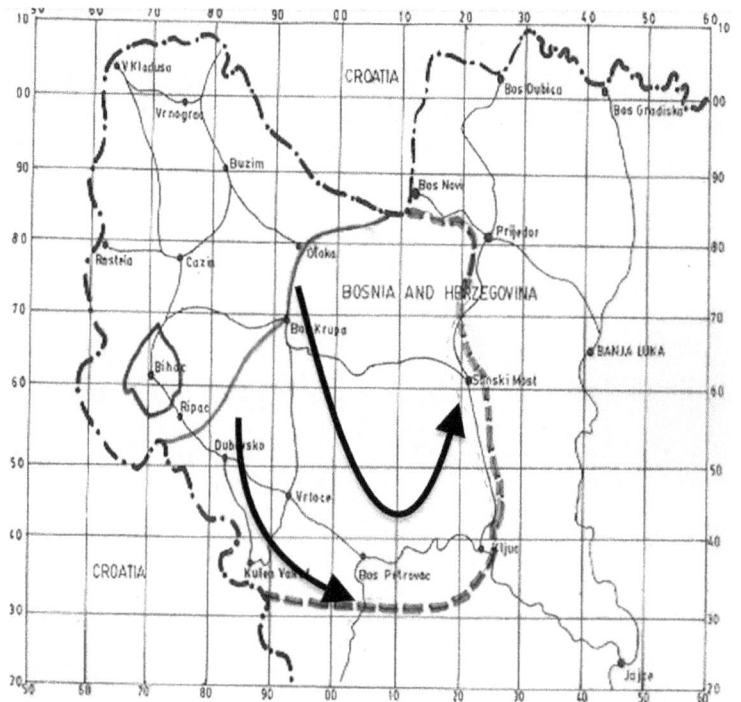

Solid black line show Bihac boundary during the war. Broken line shows the areas captured between August 1995—October 1995 (Source: Author's own wartime report records)

Media

In 1994—1995 Bihac witnessed some of the most severe fighting with ramifications on humanitarian issues. Though Bihac was a declared UN Safe Area, Serbs forces continued attacking the city and targeting civilians. Bihac would be on the 'Flash News / Breaking News' on most international news channels depicting the severe fighting and precarious humanitarian situation. Sometime in December 1994, after the repatriation of the dead bodies of two Bangladesh soldiers who had earlier died in Bihac, it created an atmosphere of extreme concern for the families in Bangladesh. Newspapers in Bangladesh had pictures of processions on Dhaka streets demanding that the government bring back the Bangladesh soldiers. Fortunately, we had an UN-provided satellite phone link to the outside world. Every day I used to get calls from Army HQ in Bangladesh for updates. I even received call from the Prime Minister and from the Leader of the Opposition at the time in Bangladesh, reassuring us that the government was working with UN New York to resolve our supply and convoy problems. These calls were indeed very reassuring that we were not a forgotten lot, at least not in Bangladesh. Copied below is a news clip from a Bangladesh newspaper of the time. The newspaper clipping is in Bengali because I wanted to use the original clip and not to use the translated version. Therefore, my apologies to readers who are not conversant with Bengali. This news clip is based on the phone interview I had with the journalists in Dhaka, Bangladesh, reassuring them that we certainly had difficulties, but were not starving and that we were fine. I could understand the journalists would never believe my reassurances and diplomatic answers. My position was, yes, we are having a difficult time, but it does not help us in keeping the families back home in Bangladesh worried and crying, which would affect the morale of the Bangladesh soldiers in Bihac. On behalf of the battalion, I wanted to reassure the families that their loved ones were safe and requested them not to be anxious.

Keeping the families in Bangladesh calm was important to keep the men away from additional worries.

Bengali newspaper clip (Source: Author's own wartime records)

Outcome of War

The long, drawn-out war in Bosnia Herzegovina did not provide the warring parties with the dream country for which they had fought the war. What each ethnic community got, in the end, was a deep scar that future generations will have to live with and the untold sufferings the leaders brought upon their people. Interestingly the Bosnian Serbs lost most of the ground they held at the start of the conflict and now hold a much smaller territory which they named as 'Republika Srpska'—Republic of the Serbs. After besieging Sarajevo, the capital of Bosnia and Herzegovina, for over three years and subjecting the city population to near strangulation, the Bosnian Serbs failed to capture Sarajevo. Bosnian Muslims were the biggest losers as the Muslim population was completely cleansed from all townships and settlements along the eastern border with Serbia and from important cities like Banja Luka, Prijedor, and Zepa. Comparatively, Bosnian Croats emerged better off from the conflict, having retained much of the area they wanted to hold Mostar and access to the port city of Split, which provides the only sea port and access to the Adriatic Sea. After three and half years of war, the warring ethnic communities were tired and exhausted. They realized that their political ambitions for a dream state was unachievable and their people had already undergone untold suffering. The Bosnian Serbs, the Bosnian Croats and the Bosnian Muslims finally signed the Dayton Peace Agreement, which resulted in the establishment of a single sovereign state to be called '**Bosnia Herzegovina**'. This new state has a complicated government structure to appease all community representation but may not be an effective government authority. The war's outcome was having Bosnia Herzegovina, a country with three distinct entities explained below and with a state structure that may not be very functional.

- The 'Bosniak-Croat Federation,' created after the Washington Agreement in 1994, ended the war between Bosnian Muslims and the Bosnian Croats.

The Bosniak-Croat Federation is the elongated central part of the country, with the area divided almost equally between the Bosniak Muslims and the Croats. Croats mostly inhabit the western part up to the Adriatic Sea coast, including the famous and historic city of Mostar. Bosniak Muslims retained the central region and areas further east, close to the Serbia border. The Bosnian Muslims mostly inhabit the capital city of Sarajevo, but much of the surrounding villages and mountains remain under Serb control.

- The 'Republika Srpska'(Serbska) is the second entity of the country Bosnia Herzegovina with Serb community and on both sides of the Bosniak-Croat Federation.
- Brcko autonomous district is the third entity of Bosnia Herzegovina. Brcko—(Berchco) has almost equal representation from all three major ethnic groups and, therefore, could not be combined with Republika Srpska or Bosniak-Croat Federation. This new self-governing district is close to the Serbia border and belongs to both entities, the Republika Srpska and the Bosnian Muslim-Croat Federation.

The international community has crafted a unique state structure for Bosnia Herzegovina as a compromise solution to bring the three ethnic communities to live together in one state. However, the complex state structure leaves doubt as to whether it will be a sustainable model in the long run. Each entity has its independent legislature leaving the central government institutions as weak entities. With a divided state structure, it remains a challenge for the country to arrive at any unified state policy. The deep scars of war may take a few generations to heal, and therefore, the country can only look upon the future with the hope that the threshold of peace will not be breached in future by past hatred and mistrust. The region's history is not very reassuring for a peaceful future. Yet, the hope is that the European nations will not allow the tragic event to reoccur again in Europe.

Copyright: GEO CONSULTS, Bangladesh

Bangladesh Battalion War Diary

File Copy - Bangladesh Battalion Commander's Diary 1994-95

MAJOR EVENTS

Ser	Date	Events
1.	18 Oct 94	BANBAT takes ops resp of the AOR Bihac.
2.	21 Oct 94	Last elms of FREBAT 3 leaves Bihac with their CO.
3.	25 Oct 94	BiH 5 Corps launches the famous Bihac offensive which results in maj gains and the world attn focuses on Bihac over coming wks.
4.	25 Oct 94	BiH captured Grabez Plateau upto Ripac (WK-7659). The BiH forces occupied BANBAT OP-13 denied access to OP.
5.	26 Oct 94	BiH forces cordoned Bos Krupa from North and South direction with an opening to the east.
6.	29 Oct 94	RSK/Abdic launched an attk from Cetingrad on area Siljkovaca near V. Kladusa.
7.	30 Oct 94	5th Corps launched an attk but could not capture Bos Krupa.
8.	31 Oct 94	BSA fired on OP-10 and 02 BANBAT soldiers were injured.
9.	01 Nov 94	5th Corps launched an attk on Bos Krupa and captured northern portion of the town.
10.	03 Nov 94	02 SA-2 missiles impacted in Bihac which injured 07 children and 25-30 public houses were seriously damaged. 04 missiles were fired on Cazin and 01 missile dropped in Buzim with no cas.
11.	09 Nov 94	02 Missiles were fired from a BSA aircraft in Bihac city on Combitex Company which injured 05 civs and caused hy material damages.
12.	13 Nov 94	BSA launched an attk on Grabez Plateau with the support of Arty, Mor and Tank.
13.	14 Nov 94	BSA resumed attk on Grabez Plateau. BSA launched 05 SA-2 missiles at Cazin.
14.	16 Nov 94	2 SA-2 missiles impacted around Cazin. ARSK and Abdic soldiers launched probing attk

through the western border of the area and
captured Bugar(west of Coralici).

15. 18 Nov 94   04 BSA Aircrafts bombed Bihac city with Napalm
and Cluster bombs. Bombs landed in HQ 5 Corps
but did not explode.

16. 19 Nov 94   2 x BSA Aircrafts (ORAO) attempted an
airstrike on Cazin at 1556 hrs and 1 x
aircraft crashed with one serb pilot killed.
15 pers were injured one later died.

17. 20 Nov 94   1 x Tank shell blew up an empty living
container in V. Kladusa.

18. 22 Nov 94   BSA/ARSK has advanced further north and
captured few croatian villages on the south of
Bihac city.

19. 12 Dec 94   One BANBAT APC from V Kladusa Log Base was
directly hit by two anti tank guided missiles
at 1518 hrs. 05 soldiers were injured, one
later died.

20. 01 Jan 95   A gen ceasefire was signed between BiH and BSA
for a period of 04 months. Result of Mr
Carters visit.

21. 14 Jan 95   HVO bde alongwith 501 bde launched offensive
in area Vedro Polje where they killed 21 ARSK
soldiers and captured 08 as POWs.

22. 27&28 Jan 95   5th Corps recaptured Bugar.

23. 10-13 Feb 95   BiH launched attks on BSA/ARSK held posns on
the outskirt of Bihac town and pushed them out
of the safe area upto present CL.

24. 06 Mar 95   5th Corps reportedly launched an attk on
Abdic/RSK occupied posns from south and east
directions. No change in CL.

25. 18 Mar 95   APWB/RSK launched attk on 5th Corps posns in
gen area Kumarica and Vrnograc.

26. 26 Mar 95   0600 hrs Abdic forces launched an attk in GA
Brezovo Polje WL-7204 where three 5th Corps
tps were killed.

27. 29 Mar 95   An ARSK/Abdic attk on 5th Corps posns in GA
Brezovo Polje WL-7304 and Hegica Glav WL-7103
was captured by Abdic/ARSK forces.

28. 31 Mar 95		An Abdic/ARSK attk on 5th Corps posns in GA Brezovo Polje WL-7304 and WL-7305. 03 KIA and 04 other injured on 5th Corps side.
29. 02 Apr 95		At 0600 hrs Abdic/ARSK forces launched an inf attk from three directions on GA Basicha BRDO WL-7203 and Brezovo Polje WL-7304 and WL-7305. The attks were repulsed by 5th corps.
30. 06 Apr 95		07 possibly tank shells and 02 other shells (type unknown) impacted inside Bihac Safe Area(town) between 1748-1752 hrs. 08 pers incl 02 children were injured and lot of material damages took place. An Abdic/ARSK attk on 505 bde posn took place. Abdic/ARSK forces captured an important hill which is locally known as Hamza Hill WK-7299. 5th Corps lost one sldr and 12 other wounded.
31. 13 Apr 95		Log Base recorded 1159 explosions and uncountable SA bursts on the east and south of the Log Base. An attk was launched by ARSK/Abdic on 5th Corps posns on the east and south of Log Base. 5th Coprs forces probably countered this attk. Abdic/ARSK forces launched an attk on 5th Corps in G. Purici WK-7094; gained some ground in Slapnica WL-7201.
32. 17 Apr 95		An attk was launched by 5th Corps on Grabez WK-7363 between 1413-1435 hrs. No change in the CL.
33. 18 Apr 95		Abdic/ARSK forces launched an attk on 5th Corps posns in GA Donja Slapnica WK-7100 at about 0630 hrs. No change in the present CL.
34. 19 Apr 95		ARSK attk in GA Velika Glav WK-7298 and Melkici Glav WK-7299 on 5th Corps posns. An Abdic/ARSK attk on 5th corps posns in GA Velika Glav WK-7298, Murtici WK-7299 and Melkici Glav Wk-7299.
35. 20 Apr 95		BANBAT Log Base recorded 499 explosions. A joint 505 bde and 506 bde attk on Abdic/ARSK posns in GA Velika Glav WK-7298 and recapture the area.
36. 24 Apr 95		01 x 120 MM Mor impacted on the grd only 90 meteres away from BANBAT Base in Bihac.
37. 25 Apr 95		13 Arty Shells impacted inside Bihac town causing minor injury to a couple of 39 years and log of material damages.

38. 28 Apr 95		07 Shells impacted in and around Bihac city centre causing lot of material damages but no cas.

Chief of Army Staff Bangladesh Army entered the pocket at 1345 hrs through Maljevac CP by rd. He was accompanied by a three member delegation. At 1630 hrs CAS awarded Medals to the BANBAT members.
08 Shells impacted inside Bihac Safe Area.

39. 29 Apr 95   BANBAT OP-32 WK-665635 in Bihac observed 02 Green Aircrafts in the southwest dir flying at a low altitude. Those aircrafts bombed GA Skocaj Wk-6856, Crnivrh WK-6159, Spasinvrh WK-6062 and Grabovorame WK-6162. Cas took place in GA Skocaj WK-6856. Not BANBAT cfm.

Joint forces of ARSK and Abdic forces attk the def posn of 505 bde at Velica Glav WK-7297.

OP-41 and OP-42 observed 13 houses being shelled by ARSK. 03 of them were burning in WK-6180 and WK-6280.

40. 02 May 95   05 Arty Shells impacted in and around Bihac hosp. A store room was partially damaged.

41. 04 May 95   03 Tank Shells landed in Bihac town injuring 06 persons. 01 woman aged 40 years died afterwards in the hosp. 02 persons were injured.

42. 05 May 95   07 Shells impacted in Bihac town injuring 02 persons with some material damages.

43. 06 May 95   02 Shells impacted in Bihac town WK-683633 injuring 02 persons.

44. 07 May 95   13 Shells impacted in Coralici. 08 Shells which are within 1000 metres of BANBAT HQ caused some material damages.

45. 08 May 95   High level activities monitored near BANBAT OP-14. Uncountable SA bursts were recorded alongwith 257 Shells. 33 explosions impacted within 1500 metres of BANBAT OP-14.

16 Shells impacted in Coralici.

46. 09 May 95   BANBAT APCs were hit by 03 rds of SA fire when escorting BAC Comd. One APC received one hit while another received the other 02 hits. No cas.

47.	12 May 95	OP-13 observed high level activities and recorded 331 explosions and uncountable SA bursts. 5th Corps has launched offensive in Buzimkici WK-8870.
48.	13 May 95	5th Corps launched offensive in Buzimkici WK-8870 and Hodzinac WK-9172. High level activities monitored near BANBAT OP-14. 5th Corps gained some grd in this front.

While escorting a convoy from Coralici to VK 01 BANBAT APC was hit by SA fire at 1015 hrs in G.Purici WK-6993. The fire came from WB forces. No cas. |
| 49. | 16 May 95 | 5th Corps launched an offensive on Buzimkici at 0945 hrs. Intense shelling and SA fire continued for 23 minutes.

BANBAT APC came under SA fire in G.Purici WK-6993 from WB forces while escorting. No cas. |
50.	17 May 95	5th corps captured Ripac town and P.Grabez in the morning. 13 shells impacted in Bihac town today killing 02 pers.
51.	23 May 95	5th Corps launched an offensive early in the morning and captured GA Hodzinac WK-9073, WK-9173, WK-9072 and WK-9172. 03 Shells impacted around Coralici air fd WK-686810. No cas.
52.	24 May 95	A POW exchange committee meeting took place in Licko Petrovo Selo between 5th Corps and RSK where 5th Corps handed over 07 dead bodies to RSK.
53.	26 May 95	08 Shells impacted in Bihac town. One man of 27 years old was badly injured.
54.	27 May 95	LO 505 Bde reported an attk was launched by joint forces of ARSK and WB in G.Gradina WK-8002, Vajagici WK-8101 and Santraci WK-8100. The attacking forces initially captured these places but afterwards 505 bde recaptured the lost grd. 900-1000 tps of WB and RSK took part in the attk. 30 ARSK sldrs were killed and many more were injured. 02 sldrs of 505 bde were killed and 09 were injured. Not BANBAT cfm.
55.	30 May 95	Log Convoy VK-22 bound for VK committed an accident while coming near Topusko. The driver of the veh injured.

56. 04 Jun 95   05 Arty Shells impacted inside Bihac town. No cas. Caused some materials damages.

57. 06 Jun 95   04 Arty Shells impacted inside Bihac town which killed one woman of 33 years and a child of 6 years on the spot and damaged some house.

58. 07 Jun 95   A dead body and POW exchange meeting was held between RSK and 5th Corps at Licko Petrovo Selo. 5th Corps handed over 39 dead bodies and 03 POWs in exchange of 04 dead bodies from RSK side.

59. 08 Jun 95   Between 1756-1800 hrs 04 x 122 MM Howitzers impacted inside Bihac town. A POW exchange meeting was held between RSK and 5th Corps at Licko Petrovo Selo. 5th Corps handed over 09 POWs in exchange of 10 POWs from RSK side.

60. 09 Jun 95   BANBAT Base in Bihac was directly targeted by BSA Arty. Between 1005-1010 hrs 05 Arty Shells impacted in very close proximity to BANBAT Base at Bihac. No damage and cas. At 1715 hrs one 130 MM Arty Shell impacted on the Gymnasium inside BANBAT Base in Bihac that damaged the eastern wall and floor of the bldg. Normally tps are playing inside. At the time of impact nobody was inside the Gymnasium therefore BANBAT escaped cas. The shells were fired from the dir between 110-141 degree.

                  Abdic/ARSK forces launched an attk on 5th corps posns on the south of Vrnograc from northeast and northwest dir.

                  5th Corps launched an attk in Buzimkici WK-8870 and WK-8970.

                  An Abdic/ARSK attk on the north of Vrnograc from three dir from Gradina WL-7703, Grosici WL-7503 and Slapnica WL-7201. Abdic/ARSK forces captured some ground.

61.   10 Jun 95 5th Corps launched an attk on the east of Grabez yesterday. Some ground gained by 5th Corps. Abdic/ARSK forces have launched two attks on Vrnograc.

62. 19 Jun 95   5th Corps has launched a limited scale attk in GA Ripac with a view to expanding their recent gains.

63. 23 Jun 95   05 Arty Shells impacted inside Bihac city. The
                     shells caused considerable material damages to
                     few house. 02 injured.

                     At about 1100 hrs a BANBAT amb UN 37230 met an
                     accident with a civ veh no BI 58658 AUDI
                     inside Bihac town. Both the vehs were rushing
                     to their own destinations due to shelling in
                     Bihac city at that time. No cas. The vehs were
                     partially damaged.

64. 28 Jun 95   14 Shells impacted inside Bihac town. A 60
                     years old woman died of shock effect due to
                     the detonation while the impacts caused
                     considerable material damages to many bldgs.

65. 01 Jul 95   02 Shells impacted inside Bihac Town. A 51
                     yrs old man lost one of his arms while a lady
                     and a child received minor injuries.

66. 04 Jul 95   12 Arty Shells reportedly impacted inside the
                     Safe Area. The impacts caused minor injury to
                     01 male of 45 yrs and considerable material
                     damages to many bldgs.

67. 05 Jul 95   An air strike on Hydro-electric plant in area
                     Kostela by a grey colour unidentified AC made
                     the sit tense in the south. The air strike may
                     be a Serb retaliation due to 5th Corps attk on
                     the East of Buzim.

68. 07 Jul 95   02 SA-2 missiles launched by BSA impacted
                     around Buzim. Both the msls were airbursts
                     causing material damages to nearby bldgs. No
                     cas rpt.

69. 08 Jul 95   Two small scale attks were reportedly launched
                     in the morning in area BUZIMKICI WK-8870 & WK-
                     8970 and in Grabez Plateau. Buzim experienced
                     another msl attk at about 2200 hrs last night
                     with no cas rpt.

70. 11 Jul 95   5th Corps launched an attk in area BUZIMKICI
                     wk-8870 at last night. BUZIMKICI is reportedly
                     captured by 5th Corps. 5th Corps claimed 50
                     BSA sldrs to be KIA while accepted 08 KIA and
                     13 injured on their side.

                     02 Arty shells impacted at IZACIC wk-6169
                     which killed a boy of 15 yrs named DIZDAREVIC.

71. 15 Jul 95   16 Shells fired by BSA reportedly impacted
                     inside Bihac safe area. The impacts killed 01
                     male of 50 yrs and injured 03 other pers

including damages of few houses and shops i:
the safe area.

72. 17 Jul 95   ARSK launched 04 SA-2 and SA-6 msls from th
Western border in the pocket. 6/7 per:
received minor injury while the nearby house!
were damaged.

73. 19 Jul 95   WB/ARSK launched a maj attk in GA STURLI(
between 0310-0400 hrs. Log Base recorde(
uncountable shells and SA fire. BANBAT OP-4(
had to withdraw under continuous ARSI
shelling. One of the APCs was hit by ;
splinter. OP-40 occupied another posn behin(
the original one to monitor the sit. One ol(
woman was killed while two other were injure(
due to the shelling at IZACIC.

74. 20 Jul 95   Attk and counter attk reportedly conducte(
both by ARSK and 5th Corps in GA TRAZAC WK-
6284. 05 Shells impacted on the outskirt o|
Cazin City WK-7580. One woman and a child wa:
reportedly injured. One house was seriousl}
damaged as one of the shells dropped on the
roof.

75. 21 Jul 95   Attk and counter attk reportedly conducte(
both by ARSK and 5th Corps in GA TRAZAC WK-
6284 and BUGAR WK-6177.

76. 23 Jul 95   WB/ARSK attks from North and West were
supported by coordinated BSA attk from the
east and southeast. 5th Corps has lost
considerable grd on the north of TRZACK/
RASTELA.

77. 25 Jul 95   Two small scale attks were launched in the
north and south. Coralici was intensively
shelled from ARSK throughout the night as
because a fixed wing ac landed in the Coralici
airfd carrying huge amount of supplies for 5th
Corps. The ac flew back safely.

78. 26 Jul 95   A WB/ARSK attk on the south of BANBAT Log Base
at WK-6582 and WK-6988. 5th Corps lost ground
on the north. ARSK launched attk on the West
at PJANICI WK-6481. 653 shells were recorded.
04 tank shells impacted within 100-300 M of
OP-41 and OP-42. 07 Rockets impacted very
close to Coralici camp within 100-200 M
causing considerable damages to many bldgs.

79. 27 Jul 95   Two small scale attks were reported from GA
BUGAR WK-6177.

80. 28 Jul 95   A maj scale attk was launched by ARSK at about
                  2350 hrs. Another ARSK small scale attk
                  reported between 0524-0532 hrs. ARSK has
                  captured part of GA BUGAR WK-6177 on the west.

81. 29 Jul 95   An attk was reportedly launched in GA BUGAR
                  and RASTELA between 0001-0220 hrs. 5th Corps
                  recaptured BUGAR from ARSK. Approx 1000 ARSK
                  tps withdrew from WB.

82. 30 Jul 95   An attk was reportedly launched around GA
                  TODOROVO WK-7393.

83. 31 Jul 95   A maj attk was launched on the north around GA
                  TODOROVO by 5th Corps on WB posns. Log Base
                  recorded 1521 shells and uncountable SA
                  bursts. Another 5th Corps attk was launched at
                  GA CAJICI WK-6785.

84. 01 Aug 95   An attk was launched in GA TRAZACK initiated
                  by 5th Corps. Another small scale attk was
                  launched on the northeast of the Log Base
                  around GA GOLUBOVICI.

85. 03 Aug 95   A maj attk was launched in GA TRAZACK. Intense
                  SA fire was recorded between 2200-0300 hrs.
                  Huge amount of ARSK tps and guns including few
                  tanks withdrew from WB. Croatia started
                  launching full scale maj offensives from
                  different direction on RSK.

86. 05 Aug 95   At about 0445 hrs 5th Corps broke out from the
                  West to link up with the Croatian Army. 5th
                  Corps captured upto GA WK-5571, WK-5667, WK-
                  5671, WK-5467 and WK-5864 on the west across
                  the boundry and ran over few POLBAT OPs. They
                  have reportedly reached GA RAKOVICA WK-5182.

87. 06 AUG 95   At about 0745 hrs 5th Corps has linked up with
                  Croatian Army in GA RAKOVICA WK-5180. Comd 5th
                  Corps met the advancing Croatian General in
                  this area.

88. 07 Aug 95   At about 1600 hrs Velika Kladusa (so called
                  western Bosnia) fell to 5th Corps. At about
                  1525 hrs adv elm of 5th corps alongwith GEN
                  DUDAKOVIC the 5th corps comd reached Velika
                  Kladusa.

89. 08 Aug 95   At about 2000 hrs last night BSA tried to
                  launch an attk in GA KLANAC WK-7656. 5th corps
                  repulsed the attk and there is no change in
                  the CL.

90.	19 Aug 95	BANBAT ptl reported total 97 Shells and uncountable SA fire in the dir of Ripac. Possibly 5th corps launched some offensive to gain more grounds towards south. BANBAT recorded 358 shells impact from the dir of Bos Krupa. 5Th corps possibly launched some offensive to regain bos krupa.
91.	20 Aug 95	Two attks were launched in Ripac and Grabez Plateau simultaneously between 1600-1700 hrs. 5Th corps may be the initiator.
92.	21 Aug 95	BANBAT recorded 11 impacts around Cazin. 05 Pers were injured.
93.	30 Aug 95	5th corps launhced an attk in GA Ripac. BANBAT base in Bihac observed red alert state throughout the day expecting a BSA retaliation to NATO Air strike near Sarajevo.
94.	01 Sep 95	5th corps launhced attks on GA Ripac and Grabez Plateua between 0808-0845 hrs.
95.	03 Sep 95	5th corps launched attks on GA Turski Kozjen WK-7862 and Tihotina wk-7759.
96.	06 Sep 95	5th corps launched an attk in GA Turski Kozjen WK-7762. Result not cfm.
97.	08 Sep 95	A 5th corps attk was launched in GA Turski Kozjen WK-7862 between 1454-1531 hrs.
98.	09 Sep 95	5th corps launched attks in GA P. Grabez, Z. Grabez, Drenovo Tizesno and Ripac.
99.	13 Sep 95	5th corps launched attks in GA Ripac, Grabez and Bos Krupa.
100.	14 Sep 95	5th corps continued offensive on the south. Another attk was launched in Bos Kurpa by 5th corps.
101.	15 Sep 95	LO 5th corps claimed that BiH has reached upto Bosanski Petrovac on the Southeast of Bihac. The ultimate aim of this advance is to link up with the BiH column advancing from Donja Vakuf.
102.	16 Sep 95	5th corps launched an attk in Bos Krupa.
103.	16 Sep 95	5th Corps captured Klujc.
104.	18 Sep 95	5th corps captured Bos Krupa.
105.	09 Oct 95	5th Corps captured Sanski Most.

Bosnia Peace Keeping Operations—Hindsight

ICRC Prisoner & Dead Body Exchange Meetings

In Bihac, the Bangladesh Battalion assisted ICRC *(International Committee for Red Cross)* in setting up meetings between the Bosnian Army 5 Corps and Krajina Serbs to exchange prisoners and dead bodies. The Bangladesh Battalion secured the tented meeting venue, which was always set up on the border on the zero line as none of the warring parties trusted each other and would not step into the other's territory. Our support to ICRC required us to set up meeting tents with generators for power and to provide protection by placing APCs and soldiers at a distance so as not to appear to be watching or hearing what was happening in the secret meetings. The two warring parties would meet inside the tent on neutral ground. During the meetings, the ICRC representatives stood outside the meeting tents; they were just the facilitators. Bosnian Army 5 Corps and Krajina Serbs representatives discussed and agreed on their secret deals, mostly about the exchange of dead bodies in exchange for prisoners. We learned later that after each battle, effort by both the warring parties was to recover and hold on to enemy dead bodies, which they exchanged to get their prisoners back. The actual exchange was done clandestinely at a mutually agreed secret place, not in the sights of the UN representatives. We never witnessed the actual exchange of prisoners or dead bodies. Mentioned below are just two of the many such meetings.

- **May 24th, 1995** meeting at Licko Petrovo Selo crossing on border NW of Bihac city. Reportedly Bosnian Army 5 Corps agreed to hand over seven Serb dead bodies in exchange for an unconfirmed number of their prisoners.
- **June 7th, 1995** meeting at Licko Petrovo Selo crossing on border NW of Bihac city. Reportedly Bosnian Army 5 Corps agreed to hand over 39 Serb dead bodies in exchange for 10 of their prisoners.

French Army High Altitude Night Drops over Bihac

During our overlap period with the French Battalion in October 1994, few Bangladesh officers accompanied the French team for two nights to a DZ *(Drop Zone)* in Bihac. We were told the air drops involved humanitarian relief supplies for the local population. For some reason, we never got to see these supplies, which we were told were handed over to UNHCR representatives on the site the same night. Since the Bosnian Army 5 Corps attack started only three days after the French left, we got busy with fast-developing battlefield events and lost track of the airdrop issue. Over the next fourteen months after the French left I assume such air drops were not done because we were never asked to pick up such air drop supplies again. During the rest of my time in Bihac, I was so preoccupied with the ongoing war events, I totally forgot about the airdrop issue. The question still remains in my mind even today, what was the real content and purpose of the French air drops at night? Why the contents were not shown or revealed to the Bangladesh military officers present on site? Why did such airdrops cease after the departure of the French? These remains as unanswered questions and I prefer to leave it for history to reveal.

Meeting Venue on the Confrontation Line (CL)

After the Dayton Peace Agreement was signed in 1995 by the warring parties, an important task given to the UN Peacekeepers was to get verified and signed maps from both the Serbs and the Bosnian Army. In the negotiations, it would be normal for warring parties to claim more land than they held. The signed maps were a declaration by the warring sides as to the actual line of control, when the ceasefire came into effect. These signed maps showed the territory the warring parties would hold within their new state boundary.

Author, far right, in confrontation line (CL) waiting to start meeting of Bosnian Government: Bosnian Serbs to sign the CL maps required by Dayton Agreement, Oct. 1995 (Source: Author's own wartime records)

It was my responsibility to arrange joint meetings in Bihac area for the warring parties to collect signed maps showing the lines they were holding, and this had to be confirmed by the other warring side. These maps were the basis for delineating the internal boundaries between three factions in the new Federation of Bosnia Herzegovina. The first meeting I arranged was about a two-hour drive away, southeast of Bihac city in Bosnian Army 5 Corp's in newly captured territory. We had to carry everything that was required to set up the meeting, which occurred on the confrontation line between the Bosnian Army 5 Corps and the Bosnian Serbs.

Author, far center, chairing the meeting of Bosnians. Government delegation on the left, Bosnian Serb delegation on the right, to sign the CL maps required by Dayton Agreement, Oct. 1995 (Source: Author's own wartime records)

Scanning the area before the meeting. Picture shows one buried anti-tank mine, Oct. 1995 (Source: Author's own wartime records)

At the meeting location, the road was narrow, with no space for the container-carrying truck to turn around, and we were not allowed to drive into the Serb territory in order to turn around. The flatbed self-loading container truck had to go in reverse to the meeting point on the confrontation line, and after dropping the container, the truck would drive back to the Bosnian side. Two APCs would then move into position to shield the meeting container from fire from nearby hills. We also had to carry a towed generator to provide light and heat inside the meeting container. We even carried coffee and soft drinks to keep the delegates from the two sides in good humor. On this first meeting, after welcoming the delegations, I requested that the two sides shake hands, a symbolic gesture to show the war was over. Though initially very reluctant, they eventually shook hands, which helped ease the meeting environment.

I had arranged and chaired the second meeting on the confrontation line right on the road south of Bihac city, past Ripac. Unlike the previous meeting place, the approach road was long, so going in reverse was difficult. The meeting had to be set up right on the single-lane road itself because both sides of the road were mined. It was too risky to attempt to demine a safe patch outside the tarmac as the area was infested with unmarked mines. The solution was placing one US-built M113 tracked APC on the Serb side of the meeting container with onboard machine guns on watch. After that, the container-carrying flatbed truck would do a long reverse drive into the position, dropping the container and returning to the Bosnian side. After connecting the towed generator for heating, the second APC was then placed for protection.

Interestingly, the Serbian delegation members at this meeting were more friendly. They even offered us the local Serb wine, which was more like Vodka. After the meeting, which went very well, the entire setup process was repeated in reverse order so that we could retrieve the equipment. We used the M113 tracked APC on the Serb side because it could pivot around 360 degrees like a tank on the road itself. This pivoting capability of the M113 tracked APC was indeed very helpful.

Support to UNHCR Humanitarian Efforts

In addition to the typical peacekeeping duties, the Bangladesh Battalion did a range of other duties to support UNHCR and ICRC. In Bangladesh, the army's involvement in counter-insurgency operations has given its soldiers good experience in community relationship building. This experience was put to good use in 1995 when the battle scenario changed, and Bangladesh soldiers had more opportunities to interact with the local village people. For example, in the winter, we used our trucks to sprinkle salt on the village roads and also cleared the snow from the roads with snow ploughs attached to our trucks. This was a service which, in normal times, would have been done by the local government. At times we sent out patrols with food packets for distribution to more remote villages. Contents of the food packets were savings from soldiers' regular food supply. I also remember assisting in repairing Bihac City High school, which was damaged by the Serb shelling. The Bangladesh Battalion had a medical unit and a dental unit at Corallici Camp, where we provided free dental service for locals. They could also come to see the doctor, and we had the clinic open almost every morning. Medicines given to the locals were from the stock of supplies we had carried from Bangladesh. This kind of community relationship-building service is typical with all Bangladesh Battalions deployed in Peacekeeping Operations in other places like Africa, where the Bangladesh Army has a major involvement.

Discord on Bihac Evacuation Plan

Around Nov 1994, the ground situation was critical and threats to UN Peacekeepers increased. The warring parties were showing no interest in peace talks, so the UN threatened to pull out all peacekeepers; the thought was to let the fire burn itself out. We received instructions to plan for withdrawal under adverse conditions as a contingency plan. Bihac was an isolated enclave surrounded by unfriendly Serb territory. Our way out would be to drive approximately 100 km through the territory under the control of Croatian Serbs. The US Navy Adriatic fleet was tasked to plan for the evacuation of the Bangladesh battalion. A US Marine officer from the Adriatic fleet was in contact with me for planning the evacuation. We had a few meetings in Zagreb UNPF HQ. The plan was to use the US Navy's large helicopters to lift us out of Bihac in the case of forced evacuation. The US Marine officer had requested a reconnaissance trip into Bihac to select the 'Landing Zone—LZ', but he did not receive permission on the pretext that he was not part of UNPROFOR. Eventually, on his request, I provided him with video footage of possible landing sites for the helicopter fleet. At another meeting on withdrawal planning at HQ UNPF Zagreb, I was briefing the UNPF operations staff on the situation in Bihac. I was surprised that the US

Marine officer was denied permission to attend the UNPF briefing again on the grounds that he was not part of the UNPROFOR staff. I could sense in UNPROFOR HQ there was a lack of support for US involvement.

Living Through Battlefield Stress

While most military professionals must have heard of the term 'battle stress', few would have experienced it. It's a terrified or confused mental state of a soldier exposed to numerous bomb explosions in close proximity. While detonations can cause death or injury, the fear of death can induce severe stress resulting in mental breakdown, and thus the person takes inappropriate decisions. While trying to terrorize the local population in Bihac, Serbs intentionally or unintentionally targeted the Bangladesh Battalion soldiers in blue helmets. Serbs used all kinds of weapons at their disposal. Besides the main tank gun and artillery guns, they frequently used heavy caliber truck-mounted machine guns on ground targets. We also experienced some unusually large detonations, which we suspect came from SA-2 air defense missiles fired on ground roles with proximity fuses which caused detonation in the air just above the ground surface. These missiles served more as weapons of terror, but did little harm to people not in close proximity.

With UNPROFOR's inability to provide us with logistic support for over ten weeks, within the Battalion, questions were asked, *'are we the forgotten lot of UNPROFOR'*. There were a few occasions when our soldiers at the OPs *(observation posts)* were subjected to targeted firing or shelling, resulting in some injuries. Bangladesh soldiers took protection and kept reporting and did not abandon their positions.

Bangladesh Battalion soldiers were exposed to this hostile environment for fourteen long months with little contact with families except for occasional letters from the UN postal system. Long exposure to hostile shelling combined with exposure to the elements of the battlefield did result in visible signs of battlefield stress on some of my officers and soldiers. This small group showed signs of fear and anxiety and expressed the feeling that the job was becoming too dangerous. They requested me to look at the possibility of withdrawing the Bangladesh soldiers from Bihac. I simply told these people that soldiering was a profession they chose voluntarily, and being in harm's way was an occupational hazard, so there was no option of pulling out. I would show up at every location where soldiers felt insecure and in danger, giving my soldiers a message that I was facing as much of a hostile environment as the soldiers faced. My men knew that my vehicle had taken hostile bullet hits on at least two occasions. This was not the first time I had been shot; I survived each time because I was in the prayers of my parents and well-wishers.

Culture & Language

From the French Battalion, we had inherited a good number of Bosnian female workers who provided cooking, cleaning, and barber services for the soldiers on UN contracts. For Bangladeshi soldiers to find females working in soldiers' accommodations and cook houses or providing haircuts was a cultural shock. The Bangladesh Battalion had its own cooks, cleaners and even barbers and focused on meeting the soldier's needs. Bosnian female workers became very worried that they might lose their jobs with the arrival of the Bangladesh Battalion, and I did not want that to happen. It was a survival need for these local workers to continue to support their families, because the males had all been drafted into the military for fighting. We needed time for the soldiers to adjust, so I sent the Bosnian female workers on leave with the assurance that they would not lose their jobs. During that time, we carried out education and motivation for our soldiers not to question the way of life of the local community. Happily, the soldiers understood and adjusted to the new environment. The female workers were back on the job, and we had no complaints.

There was another cultural shock experienced by the Bangladesh army officers during their overlap stay with the French. The shower enclosures did not have any screen or door. The French officers would normally walk around fully bare-bodied. To Bangladesh people, it was a shock to see a nude person walking amongst them. Bangladesh people would not strip naked in front of others which was common practice within the European shared accommodation. To overcome the problem, our people had to adjust their timing so as to avoid encountering the French. Later after the French departed, we installed curtains and doors for privacy.

In the operational area, communicating with the locals or at the checkpoints required interpreters. We had noted this difficulty during our reconnaissance the year before and in discussions with the French. I discovered that in the Bangladesh Navy, there were officers who had done years of training with the former Yugoslav Navy earlier and could speak the Serbo-Croat language. Bangladesh Navy provided four such officers who were good at speaking Serbo-Croat. I remember my first trip to Bihac along with a Danish military officer; we were stopped at the Serb checkpoint when my naval interpreter, Lt. Cdr Rouf, started speaking in Serbo-Croat with the Serb soldier who had his eyes popping out in shock and surprise. The Serb soldier became more friendly, and we were allowed to proceed. As we moved on, the Danish Officer commented, *'Salim, you have a secret weapon which should prove very helpful.'*

As we experienced later, the Bangladesh naval officers proved extremely useful in meetings and at checkpoints. I had assigned three of our naval interpreter officers to support operations within Bihac and one naval interpreter, Lt. Cdr Habib, was assigned to stay with UN HQ Sector-North in Topusco in the Croatian Serb area. The job of the Lt. Cdr Habib was to act as a liaison with RSK , Croatian Serbs, to coordinate and get approval for our convoy passage clearances from the Krajina Serbs at Knin. Habib soon developed good contacts with the Krajina Serbs, which helped resolve many travel and operational issues. For all my meetings with the Bosnian Army 5 Corps, I always had my naval interpreter Lt. Cdr Rouf with me. These four naval interpreter officers proved to be of immense value to us during our long stay as they could easily get into talking terms with the warring parties. An important lesson to be taken is that talking in the local language helps ease the tense atmosphere in a conflict situation.

Winter Training and Winter Survival

On arrival at Zagreb and to make use of the available time before operational deployment to Bihac, we planned for winter training as much as we could. Bangladesh soldiers could easily be identified in Camp Pleso, Zagreb, with their heads and faces covered in all kinds of headdresses to protect against cold. For training on winter survival, we had assistance from the US Army Doctors from MASH located in Zagreb. US medics took classes on 'winter adaptation and winter survival' techniques. Bangladesh officers who attended the classes repeated the class content in classes for the soldiers to pass on the acquired knowledge. Training included care and use of thermals, including sleeping bags which were not a standard issue item for soldiers in Bangladesh. Bangladeshis have a habit of rubbing oil on the skin after a bath, and the US medics explained that oil degrades the skin insulation, so this practice was to be discontinued. This was a practice the soldiers had learnt from childhood and was difficult to stop. The importance of maintaining body core temperature was another interesting subject. As we learnt, simply drinking hot water was better than drinking tea or coffee or any drink with caffeine. On the operational side, we had a few critical issues on training like radio set batteries upkeep in low temperatures, winterization of vehicles and snow driving and learning to watch out for black ice. Unfortunately, we could not execute the winterization program for all APCs and soft vehicles as fighting broke out in Bihac soon after our arrival. Serbs imposed blockade prevented the UN technical support team from visiting Bihac for vehicle winterization. The use of snow chains was a new experience for our drivers, who had never seen snow before and had never used anything like the snow chain.

In the absence of UN technical support, we had to do a few of our own improvisations for survival. We were using water in APC radiators, and due to the icing of the radiator water the radiators could crack. We did not have coolants in those days. To prevent such damage, I introduced the practice of keeping the radiator uncapped for the night, which allowed the expanding iced water to escape instead of damaging the radiator. In the morning, one could see a white flower-like ice candy on top of each radiator which would melt away with daylight or when the engine started. BTR 70 APCs had poor-quality batteries; we faced the problem of APC batteries dying out on winter nights. We did not face the battery problem with other vehicles, which were mostly of German origin. To mitigate this issue, I introduced the practice for the APC drivers to do static engine runs at midnight for a good half an hour to fully charge the batteries and to keep them warm; this worked but was a tough call for the APC drivers. This became a mandatory duty for APC drivers before going to sleep.

On the operational side, during the winters when the BTR 70s APC were sent on night watch patrol, after remaining static few hours, the wheels would sink in the snow, and the APC tires would start slipping. We looked for snow chains but were told that the Russians never had snow chains for BTRs. My immediate solution was to ask the crews to deflate the tires until they looked flat and round, and this increased the tire contact area and the traction. With deflated tires, we never again faced the problem of BTRs getting stuck in the snow. I was happy all my improvisation ideas were working well.

Winter driving was always a big concern because, in Bosnia, there were more peacekeeper fatalities due to road accidents during the winters than war-related causes. We all had to learn about the dreaded black ice and how and where to watch for it. Our driving technique insisted on avoiding the use of brakes, especially when going downhill.

We taught drivers to drive in low gears, sometimes on first gear, while going down the slope. Fortunately, all our vehicles had manual gears, so switching to low gears was an easy and safe technique. At the end of the mission, I looked back; we had only two snow-related accidents with minor injuries. This was insignificant compared to other nation s in the mission area. Maybe we overdosed, and the drivers were cautious, which saved the battalion. The most notable fatal road accident of the Bosnian war occurred in Aug 1995 when some senior American diplomats, along with a French soldier, were killed while travelling in a French Army APC as it slipped and went down a deep ravine on Mt. Igman road near Sarajevo.

Our soldiers' accommodation was in prefab containers which required heating during the winters; therefore, there was high electricity demand. The electricity demand by the UN camps was met by running generators which again required large fuel stocks. During the Serb-imposed blockade of ten weeks from late October 1994 to December 1994, we had to survive on fuel that was in stock with no assurances of resupply. Fuel had to be rationed; therefore, generator hours had to be restricted. As part of rationing, we stopped providing electricity for heating at night after 11 pm. Warm clothing and sleeping bags provided warm comfort. We had seven 100 KVA generators running in our main camp, and the generators required weekly topping up of engine oil. We came to a point where we had no engine oil for the generators. Switching off the generators was not an option. So, when the problem was presented to me, people were shocked by my solution. I instructed the technicians to drain out engine oil from the unserviceable APCs and filter the engine oil using surgical bandages taken from the medical unit in layers. This filtered oil was then used for topping up oil levels in the generators. This improvisation worked very well to keep the generator running to provide the minimum need for heat and electricity.

The core element of the Bangladesh Battalion was 10 East Bengal, a non-mechanized infantry battalion. We were given around seventy BTR 70 APCs in the mission area. Our APC drivers were essentially truck drivers who received only four weeks of basic driving training in Slovakia. The crews certainly required more practice to improve their driving skills, especially for mountainous areas. We deployed the crews with no training on 'Tactical Combat Driving', fortunately these inexperienced crews did manage to accomplish the tasks given.

Driving Under Hostile Fire

On December 26th, 1994, I had scheduled a meeting with Fikret Abdic, the muslim breakaway faction leader, to discuss UN proposed ceasefire. The meeting required me to drive across the confrontation line between Bosnian Army 5 Corps soldiers and Serb soldiers in support of Fikret Abdic's forces. There had been severe fighting only a week before, and the confrontation line was not fully settled. I anticipated hostile fire during the crossing of the confrontation line as soldiers would still be jittery, the tendency being to fire on anything that moved. I was not sure how the APC drivers would react under fire, so I preferred to self-drive a British Army armored Land Rover instead of using the APC. I had to carry my interpreter, a local employee, and one UN Civilian Affairs officer, a US national, in the Land Rover. It proved to be a very good decision because on the return trip, after sunset, and as we were crossing the confrontation line, some sleepy soldier on the front line started firing, which triggered firing from both sides across the entire confrontation line. I pushed my gas pedal down to the max to cross over as quickly as possible. I guess the exposure may have been for only a few minutes, but I could clearly hear bullets hitting the Land Rover body. Fortunately, none of the bullets hit the tires , which could have caused a fatal accident sending my vehicle down the ravines. After I reached my camp safely, I checked the Land Rover body and counted six hits. I have to say the British Army armored Land Rover is a good product which saved my life.

Home Leave

In Bosnia Herzegovina, the war conditions and frequent threats to one's life created a high-stress environment for the UN Peacekeepers. Most nationalities practiced sending their soldiers on home leave. In Bihac, stress was more due to a lack of freedom of movement, isolation from the outside world, and little contact with families and loved ones. Life was reduced to a routine of daytime battle gears for duty, meals and sleeping containers. We were completely cut off from the social world for months at a time together. I could see isolation and battlefield stress's visible effects on officers and soldiers. To relieve the effects of isolation, I planned to send my soldiers on home leave to Bangladesh as part of mental recovery. Due to the long flight distance, the air tickets were quite expensive on a commercial flight. So, my decision on home leave surprised many people because the Bangladesh Army soldiers elsewhere on peacekeeping missions did not practice home leave. I went ahead with my decision and designated Lt. Col Sayeedur Rahman to negotiate for charter flights which he did successfully and at a fairly cheap price. We were getting Boeing 707, an old-time jet; the aircraft choice was to minimize cost. We planned the home leave in batches of around 180 people; the figure was limited by aircraft capacity. To further reduce costs, on the flight plan, we avoided using large commercial airports for departure and mid-journey refueling breaks.

Mission area leave policy allowed 21 days of absence from the mission area. Our contract was for seven chartered flights. The plan for the first flight was to go to Dhaka, drop off the first leave batch and return empty. Subsequent flights were scheduled after every 20 days. The flight would go with the next outgoing leave batch, drop them off at Dhaka and bring back the returning leave group. The last flight had to go empty and return with the last leave group. I can recollect each person paid USD 640 for a two-way journey, which was half the cost of a ticket on a commercial flight from Zagreb to Dhaka by KLM, which was USD 1,200 at the time. The critical issue in executing the leave plan was to get each group of 180 men out of Bihac, passing through the Serb-held area into Zagreb a day before the flight. It was before a scheduled flight departure date; fighting flared up after Croatian Serbs fired some Scud missiles at Zagreb. This resulted in heightened tensions, and all UN convoy movements were stopped. We were lucky to have extricated our batch of 180 soldiers out to Zagreb in time to catch the flight, failing which we would have lost a lot of money paying compensation for the chartered flight.

Afterthoughts

Peacekeepers without a UN Mandate

Peacekeeping Operations are conducted based on a UN Security Council (UN SCR) mandate. The precondition for such a mandate and peacekeeping operation is that the belligerent parties agree to a ceasefire. The parties may have demands, some of which may not be agreeable to the other side and may require prolonged negotiations before finding a middle ground. Ideally, a peacekeeping force is deployed after the ceasefire, and the UN Security Council (UN SCR) resolution is agreed upon. Peacekeepers are not expected to fight, but can fire in self-defense, which would mean when under direct threat or when there is a danger to life. The problem with peacekeeping operations in Bosnia Herzegovina was the absence of an agreement between the Serbs, Croats, and Bosnian Muslims to a ceasefire and, more importantly, the absence of the UN Security Council (UN SCR) mandate for peacekeeping. With no signs of the war stopping, ongoing fighting had already forced a significant part of the population to move out of their homes and become refugees. In the absence of any UN mandate for peacekeeping, the only thing the Security Council could agree on regarding peacekeepers at that time, was to ask them to escort UNHCR Humanitarian Relief Effort convoys. The brutal war continued for over three years in the presence of the Peacekeepers because the big powers at the UN Security Council could not agree to a common viewpoint on when and how to stop the fighting. Clearly, the lead nations on the council had conflicting interests in Bosnia Herzegovina and therefore failed to find common ground to enforce a ceasefire. One may ask whether NATO and the EU together could have prevented or stopped this atrocious war; the answer is YES, they certainly could have stopped it much earlier, something they finally did towards the end of 1995. After three years of indecision, NATO eventually launched its air campaign on August 30th,1995, which signaled to the warring parties when and where to draw the lines and to stop fighting.

Towards the end of the war, the joint Croat and Bosnian Government forces, had captured the town of Sanski-Most, threatening the Serb stronghold of Prijedor and Banja Luka. This was not a desirable outcome for the war, as it would weaken the future Bosnian Serb state. I believe the Bosnian Army was warned and told they would be bombed if they crossed the line to go any further. So, the Europeans were balancing out the three ethnic communities on their future borders for the new state. Finally, in November 1995, under international pressure, Serbs, Croats and Bosnians, all agreed to come to the negotiating table with the peace agreement being signed on December 14th, 1995, in Dayton, Ohio, USA. Had the Serbs lost Banja Luka, their position in the new emerging confederation of Bosnia Herzegovina would have become weak. Certainly, the lead European nations did not want that to happen. So now they had a country with a balance of power and territory. The new country that emerged was a complex federation model, with a rotating Presidency. Decision-making in the new state of Bosnia Herzegovina can be complex and difficult. Only time would tell whether Bosnia Herzegovina would be a viable state in its current form.

Why the War in Bosnia Herzegovina continued for over Three Years

If the war in Bosnia Herzegovina had not continued for such a long period, the scale of atrocities committed, and the killings would have been much less. So many people did not have to die to satisfy the political objectives of a few. So rightfully, one may ask today why the war continued for over three years in modern-day Europe. Could the EU, NATO or the UN not stop the war anytime earlier? Having attended multiple operational meetings and reviewing various reports and documents during the wartime, it became apparent to me that the leading European powers were deeply worried about the emerging country's shape and character. According to their perception, Bosnia Herzegovina, with its predominantly Islamic population, could potentially support the jihadist movement and become a future threat to Europe. Additionally, with most of the land under Serb control, the Krajina Serbs could be armed and supported to secede from Croatia, leading to destabilization. Furthermore, if much of the Bosnian territory remained under Serb control, it would prevent Bosnia Herzegovina from functioning as an independent state. In my opinion, the war was allowed to continue to achieve two political objectives stated below, and the peace talks were merely a smokescreen to buy time to achieve the desired outcome.

- First—To let the ethnic communities carve out territories to the extent that no single group emerged stronger than the other two.
- Second—The new nation of Bosnia Herzegovina should not have the character of an Islamic state.

After the breakup of Yugoslavia, the most destabilizing issue was the declaration by Croatian Serbs to establish an independent and self-governing republic in southern Croatia, known as the Republic of Serb Krajina (RSK). Croatia did not allow a part of its country to secede, leading to a three-year-long unsuccessful effort by the United Nations to resolve the issue. This RSK territory shared common borders with Bosnia, which affected the stability in Bosnia. For instance, the encirclement and attacks on Bihac in northwestern Bosnia were joint operations by Bosnian Serbs and Croatian Serbs. In Bosnia, Bosnian Serbs controlled much of the land and, with military support from Serbia, became a powerful force which disregarded the UN and EU.

Continuing the war allowed Croatia to build up its military strength and launch attack in August 1995 to regain full control of its southern territory. The Croatian Serbs became refugees in Bosnia, and some went to Serbia. This also freed Bihac in northern Bosnia from three years of Serb strangulation. In addition to losing territory in southern Croatia, the Serbs lost ground in northern Bosnia and a few other places in central Bosnia, weakening their military power. Consequently, they agreed to come to the negotiating table. The Bosnian Serb leaders, President Karadzic and General Miladic, escaped and took refuge in Serbia.

When the new map, based on territories captured by each ethnic group, aligned with the European policy

```
1)We do not agree now or in the future to "arm or train" the
  Muslims within Bosnia - Hercegovina with military hardware.
2)We will continue to help impose & enforce the U.N. embargo
  on weapons to this region.While we are well aware that
```

```
  forces within the region.it is of paramount importance that
  we make sure that no such efforts are successful on behalf
  of the Muslim within the region from Islamic States &
  Groups.
  To this end & until the final outcome of the situation on
  the ground i.e. the dismemberment of Bosnia - Hercegovina &
  its destruction as a possible "ISLAMIC STATE" within Europe
  which will not be tolerated.we will continue to follow this
  policy.Further,the mistake of training & arming the Afghan
  fighters against the forces of the former USSR & their
  becoming so-called "Islamic Fighters" now in other parts of
  the world.as in Bosnia - Hercegovina,will not be repeated
  with the Muslim population in Bosnia - Hercegovina.This
  could lead to serious problems in the future within the
  emigra Muslim population within the E.C., & North America.
```

```
3)Until the situation in the former Yugoslavia is settled we
  must at all costs make sure that no state that can be
  deemed "Muslim" is allowed any say on the West's policy
  actions                                    .It is
  therefore,necessary to continue with the
          peace talks in order to delay any such possible
  actions until Bosnia - Hercegovinia no longer exists as a
  viable state & its Muslim population is totally displaced
  from its land.
```

```
  that this is infact "real-politic" and in the best interests
  of a stable Europe in the future,whose value system is and
  must remain based on a "Christian-Civilisation" & ethic.
```

Bosnia letter (Source: Author's own wartime records)

objectives, and NATO intervened with an air campaign to prevent the ethnic groups from capturing additional territory. By November 1995, the Serbs were rapidly losing ground and were on the verge of losing crucial cities such as Prijedor and Banja Luka. The Bosnian Muslims had to halt their advances when warned that NATO would bomb them if they continued.

The new state of Bosnia Herzegovina that emerged as a result of the war was no longer a Muslim-majority state. Therefore, the country no longer had the character to be deemed an Islamic state; a desired policy objective of the lead nations was thus achieved, but at a great cost to human lives. Though Europe had given super quick recognition to the declaration of independence by Bosnia Herzegovina, they did not want to see a Muslim state in heartland Europe which some thought could spread radical Islam in the region. Copied at page 247 is a letter that was in circulation at the time, and somehow, I received a copy, but I am sure I was definitely not the intended recipient. On the copied extract, I have deleted the originator's information as I cannot vouch for the document's authenticity, and it might as well be a fake document. However, the contents of the document match the outcome of the war, and therefore, the document remains of interest.

UN Peacekeepers Humiliated

Serb forces had become arrogant and cared little for the status of the UN Forces. On numerous occasions, especially in central Bosnia, Serbs took UN Soldiers hostage to negotiate their demands. This clearly showed the total helpless state of the UN Peacekeepers in Bosnia Herzegovina. I remember on my first visit to Bihac in 1993, at a Serb check post, I noticed one French Armored Personal Carrier (APC) in white color in possession of RSK Serbs soldiers. On my query, the French officer told me that the Serbs had taken the APC from the French soldiers by force. I fail to understand why soldiers from a well-trained and well-equipped army would surrender equipment as big as an APC to local Serb militia. Failure of the UN Peacekeepers even to protect themselves in Bosnia Herzegovina, and the inability of the European nations and the UNPROFOR Commanders to take appropriate decisions to stop the genocide, was tragic. It will haunt the international community for a long time to come. I also suspect the political leadership may have left the on-ground military commanders handicapped with restrictions connected to achievement of national political objectives. Inaction by UNPROFOR military commanders became the pattern that left the Peace Keeping soldiers on the ground to the Serbs' mercy. Serbs understood well that UNPROFOR would not take any punitive action. UNPROFOR's conduct in Bosnia was a good example of how not to conduct Peace Keeping Operations. The ability to take quick and appropriate decisions through the fog of war is a rare quality not easily found in many brilliant peacetime commanders. I also noticed that some of the UNPROFOR senior military commanders were very concerned about TV crews on their itinerary to ensure they were well covered on the media channels.

There have been a number of publications and reports, including that by the UN Secretary-General on why the UN failed in Bosnia Herzegovina. Unfortunately, such reports probably remain buried deep in the UN archives. However, the mass graves in Srebrenica and other locations will haunt future generations and keep refreshing the memories of the Bosnian genocide and serve as a testament to UNPROFOR's failure in Bosnia. Peter Jennings' TV talk show aired on March 25th, 1995 on ABC TV USA graphically illustrates when and where the UNPROFOR command failed. I was interviewed as well. I had the privilege of having pre-recorded interview clips for this show. My portion of the interview was recorded in Zagreb, Croatia, since I declined to travel to another country. This TV show must have caused embarrassment for the UNPROFOR and the UN.

During my part of the interview on Peter Jennings' TV talk show I explained how UNPROFOR failed to support the twelve hundred peacekeepers from Bangladesh when their food and other logistic convoys were barred by the Krajina Serbs from reaching our location in Bihac. So, the Bangladesh Soldiers survived on pack rations for about ten weeks during the cold winters between Oct— Dec 1994. Lack of fuel for heating systems, the generators and operational vehicles created additional difficulties. More importantly, I had to mention about the failure of UN Logistics to deliver our shipped containers before we were asked to deploy into Bihac, these containers arrived more than a month after our arrival in Bihac, so we were forced to deploy without our essentials equipment and heavy weapons.

Next, I was asked what I had to say to the Serb accusation that Bangladesh soldiers had provided grenades to the Bosnian Government forces which had Arabic inscriptions. Serbs linked the Arabic inscription to Muslim origin country and since Bangladesh soldiers were Muslims so they assumed grenades must have been provided by the Bangladesh soldiers. Sometime later, the same question was asked to me by the UN TV crews. It was interesting how the different media outlets had same list of queries. To clarify, I had to show our ammunition which were product of Bangladesh Ordinance Factory and told TV interviewer that our ammunitions and grenades had the inscriptions 'BOF' *(Bangladesh Ordnance factory)* in English so the Serb assumptions and accusations were not true as none of our ammunitions had any Arabic inscriptions. I know many people in the area would still not believe in my statement which did not fit into their agenda.

Conflict of Interest in Peacekeeping

All UN peacekeepers are expected to stand in neutrality and work towards fulfilling the UN Security Council Mandate. They wear blue headgear as a symbol of their affiliation with the UN. However, the military commanders at the decision-making level from countries with global and strategic interests, while in blue headgear, may have a national agenda to pursue, which can prohibit them from taking appropriate operational decisions.

During the Bosnian War, it is suspected that the top UN military commanders were most concerned about what their own countries wanted them to do, and their national interests took precedence over the UN mandate. European nations feared that a Bosnia Herzegovina with a dominant Islamic character could breed radical jihadism in their backdoors, based on jihadist activities in other parts of the world. This perception significantly influenced UNPROFOR's operational decision-making during the Bosnian War.

Religion was Not the Cause of Conflict in Bosnia Herzegovina as Portrayed by the Media

During my fourteen-month stay in Bosnia, I had the opportunity to travel extensively and meet with the local population to gain a better understanding of the conflict and the role of Islam in Bosnian society. Despite being a muslim country, my impression was that the Bosnian people saw themselves as Europeans first and foremost and were fiercely proud of their ethnic identity, whether it was Serb, Croat or Bosnian muslim.

Their daily lives reflected European culture, and the rural areas had a lifestyle similar to that of other European villages.

During my visits to some Bosnian muslim families in the villages, I was welcomed with a glass of wine that they brewed themselves and bread baked in the wood oven in their house. I noticed that the few mosques I came across in rural areas were often locked, and upon inquiry, I learned that they were only open for Friday prayers, as most of the men were drafted into the military and were away on the battlefield.

In my judgement, the conflict in former Yugoslavia was a deeply ingrained inter-ethnic conflict, which had been brewing for centuries. Religion was used by all sides as a tool to fuel inter-ethnic animosity, as it helped to recruit more volunteers for the respective ethnic military forces. However, the portrayal of the war in Bosnia by the media, emphasizing religion as the primary cause, was not an accurate reflection of the conflict.

The war in Bosnia was triggered by the declaration of independence in March 1992, and not by any religious proclamations. The Bosnian muslims feared that Bosnia would merge with Serbia to form a mini-Yugoslav state, as the Serbs had occupied most of the important government positions in the former Yugoslavia. Furthermore, the military was entirely under Serb control, with most senior officers being Serbs only. Another factor that led to the war was the desire of the Serbian military and Bosnian Serbs to prevent the Bosnian Muslims from establishing an independent state, separate from Serbia. The Bosnian Serbs favored the unification of Bosnia Herzegovina with Serbia to realize their dream of a greater Serbia, a Slavic country.

Concept of UN-Protected Safe Areas Fails

During times of war, the establishment of UN-protected Safe Areas is crucial to prevent genocide, ethnic cleansing, and killings of unarmed civilians. However, in my experience in Bosnia, implementing Safe Areas can be very challenging. Ideally, establishing a Safe Area would require agreement from all five permanent members of the Security Council, but this may not always be feasible. Additionally, finding sufficient military resources to protect the Safe Area can be difficult. It is important to have a robust UN force in place that can serve as a credible deterrent to the warring parties, letting them know that the UN can respond if the red line is crossed.

Additionally, it is crucial to have military leadership with experience and nerves to handle the battle situation, see through the fog of war, and take timely decisions. Unfortunately, during the Bosnian war, the UNPROFOR leadership in Bosnia Herzegovina lacked the ability to foresee the intentions of the warring parties and make timely decisions. This resulted in the loss of the most important element of deterrence and repeated failures. The other reality is that the belligerent parties, in all likelihood, will have political backing from one or more of the big five powers who approve the UN Resolution to establish the Safe Area. Therefore, the big powers will possibly never agree in unison to any decision to enforce protection measures required for Safe Area.

My experience in Bosnia also tells me that Safe Area does provide an advantage to the side which is being protected. The protected population are likely to have armed elements or militias inside the Safe Area, which will remain the sticking issue with full implementation. This happened with all the declared Safe Areas in Bosnia. The belligerent party based inside the Safe Area tends to keep its forces inside the Safe Area with two purposes. First, to provide security to its people as it cannot fully trust the UN Forces as we saw in Srebrenica.

The Peacekeepers simply stood watching as Serbs rounded up the male population and later killed them. Second, the presence of armed elements inside the Safe Area presents an opportunity to drag UN Peacekeepers into the conflict. There will be situations when forces outside the safe area may fire at targets inside the safe area which may result in collateral damage to UN forces. I had noticed that the Bosnian soldiers, at times deliberately took positions very close to UN Observation Posts (OP) held by Bangladesh soldiers and sniped at Serb positions. Serb soldiers in far distance would respond by firing at Bosnian soldiers with all likelihood of collateral damage to nearby peacekeeping soldiers. We did have situations when Bangladesh soldiers were injured in such firing. When UN Forces are targeted, they can call for an air attack, and that is what the protected belligerent force inside the Safe Area wants. In our case, we could have called for an air attack on the suspected Serb positions. We refrained from escalating for minor incidents such as sniping as we did not want to be dragged as a party to the conflict. Bosnian forces would have been happy if NATO aircraft were asked to attack the Serb positions.

Srebrenica is a good example of UNPROFOR's failure to protect a Safe Area as required by UN Security Council Resolution 819 of 1993. This resolution demanded that Serb military forces withdraw and stop armed attacks against the city, but UNPROFOR failed. The UN had further directed the UNPROFOR to evacuate the sick and wounded to a safer location for treatment, but UNPROFOR was unable to do so. In my assessment, the undermentioned factors may have contributed to the failure of UNPROFOR to protect the Safe Area.

- Inadequate military resources available to protect the designated Safe Area.
- Failure of UNPROFOR Commanders at different levels to make the correct battle assessment and failure to take timely decisions.
- Air effort was available but UNPROFOR leadership were unwilling to use it for fear of the unknown.
- Unwillingness to commit the UN Forces in harm's way, thus becoming onlookers. To my mind being in harm's way is an occupational hazard for professional soldiers and has to be accepted.
- There may have been external factors like home country policy directives which may have impeded the commander's decision-making.

In May 1993, UN Security Council Resolution 824 UN declared the townships of Tuzla, Zepa, Gorazde and Bihac also as Safe Areas. This resolution demanded that Serb military forces withdraw and stop armed attacks against the town, and full access should be provided to UNHCR for humanitarian relief operations. Furthermore, in June 1993 UN Security Council Resolution 844 authorized the use of air power in and around the Safe Areas in support of UN Forces.

UNPROFOR failed to implement any of the UN Security Council Resolution decisions. NATO aircraft were always overhead on CAP missions *(combat air patrol)* and supported by AWAC aircraft flying possibly over the Adriatic Sea. What prevented the use of air power to protect the safe areas and the UN Peacekeepers when they were under direct threat will probably never be answered, at least not in the public domain. In central Bosnia, UNPROFOR did carry out a few token air strikes on abandoned tanks and positions, which were not of any military significance. Such efforts may have given a message to Serbs of UNPROFOR's unwillingness to respond, so the Serb attacks on safe areas continued in pursuance of Serbs policy to ethnically cleanse the areas.

After the war, the Hague tribunal ruling in 1995 did implicate the UN Soldiers for their part in the failure to protect unarmed civilians and to prevent the Srebrenica massacre in an UN-designated safe zone. But the tribunal did not dig into reasons for failure and did not apportion blame on any of the peacekeepers.

End of Mission & Return to Bangladesh

The Bangladesh Battalion was deployed in Bosnia under the most demanding conditions. Within a short timeframe, we were required to induct and train on a lot of new equipment, including the APC, which was a mission-critical requirement. Bangladesh soldiers were facing the European winter for the first time, and they did adjust to the cold quickly. Long exposure to hostile environments and isolation was a challenge soldiers had to cope with, and they did that successfully. It was simply a display of absolute resilience, mental strength, and a high standard of discipline. The Bangladesh Army can be proud of its soldiers' ability to withstand limitless hardships and perform as their military commanders ordered. At the war's end, Bihac was no longer an isolated enclave and had no visible threats. Therefore, before our departure UNPROFOR sent only a small detachment of British Army soldiers to take over the UN facilities and the UN property. The UN plan was for most of the usable equipment and living containers valued at hundreds of millions of dollars to be removed and stocked at Aviano Italy for future use. After handing over the equipment to the British Army, the Bangladesh Army soldiers flew back home to the great relief of their waiting families.

For Duty, For Country

One may ask what did Bangladesh Battalion achieve during its mission in Bosnia. Having to lead my soldiers over some very difficult and trying times for fourteen long months, I can say that the Bangladesh Battalion kept the UN flag flying, and the UN declared the Safe Area of Bihac remained safe and secure for the Bosnian population. One may also recall the mass killings by Serb forces after capturing the enclaves of Zepa and Sabrenecia as UN military forces withdrew back to camp safety and stood to watch. I can say with great pride that the Bangladesh battalion stood up against all odds under hostile fire, did not retreat, lived on meagre supplies, faced a ten-weeks supply blockade, and faced the European winter, something they had never seen the likes of before. Few NATO officers confided to me that most European battalions would have refused to continue as peacekeepers under the atrocious conditions that the Bangladesh Battalion faced, yet they held on to their positions. Credit goes to the absolute tenacity and courage of the Bangladesh soldiers and officers who served under my leadership and delivered what I asked for under life-threatening conditions. In the end, UNPROFOR leadership appreciated our efforts. Some of the appreciation memos I received are copied in the book for the sake of leaving them on record. Finally, I have to say that I was astonished to see the resilience, tenacity, and grit of the Bangladesh soldiers in a far away land. For the men, the motivation was the country's trust in its soldiers, and the Bangladesh Battalion in Bihac Bosnia stood tall and did not fail.

Lieutenant General R A Smith DSO, OBE, QGM
Commander, United Nations Protection Force
Sarajevo
Republic of Bosnia & Herzegovina

8225

Colonel Salim Akhtar
AHQ
QMG's Br (M&Q Dte)
Dhaka Cantonment
Dhaka
Bangladesh

17ᵗʰ November 1995

Dear Colonel,

I write at the end of your Battalion's tour to say how impressed I have been by your performance.

Yours was a long and difficult task, isolated, frequently finding yourself low on the list of priorities. Your sensible, steady and reasonable approach matched with an evident ability to comprehend the bigger picture contributed to your success. Above all I commend you for the leadership you gave to your Battalion; you bought them through many hard and sometimes dangerous times.

I regret that I have visited you and your Battalion so infrequently and I know you understand why this was so. Even at the end a more pressing operational need prevented me coming to say goodbye. Nevertheless, when I did see your soldiers I was always impressed by their high morale, good humour and professionalism. They were a credit to you and Bangladesh. I was confident Bihac Area was in safe hands.

Again, I am sorry I was not able to see you all before the end of your tour.

Yours Sincerely,

Rupert Smith

Appreciation letter from Gen. Smith

07/12 '94 22:35   ☎7241   COMCEN-BHC FWD   ☐ 0

HQ BH COMMAND FORWARD

Page 1 of 1

IMMEDIATE

UN RESTRICTED   # OUTGOING   1/1

OUTGOING FAX NO: 3383/94	DATE/TIME: 072158A DEC 94
TO: BANBAT	FROM: BH COMD FORWARD CH G3 OPS LT COL ILLINGWORTH SIGNATURE:
FAX NO:	FAX NO: VSAT 7239, CRYPTO 7240
ATTN: CO BANBAT	FILE REF NO: 3182 DRAFTER: LT COL ILLINGWORTH
INFO: FAX NO:	
SUBJECT: BANBAT WITHIN THE BIHAC POCKET	
INTERNAL DISTRIBUTION:	

MESSAGE

PERSONAL FOR COLONEL SALIM. *Dear Colonel*

I THOUGHT THAT IT WOULD BE EASIER TO COMMUNICATE ON THIS MEANS
THAN THROUGH THE PRIVATE MAIL SYSTEM. YOU MAY HAVE HAD A LONG
WAIT FOR MY LETTER OTHERWISE!

I WOULD LIKE TO PUT ON RECORD, AS A JUNIOR MEMBER OF THE STAFF
IN SARAJEVO, HOW MUCH THE EFFORTS OF YOURSELF, YOUR OFFICERS AND
MEN ARE APPRECIATED HERE WITHIN THE HEADQUARTERS. WE CAN ONLY
STAND BACK AND WONDER AT YOUR RESOLVE, PATIENCE AND DEPTH OF
CHARACTER IN THE WAY YOU HAVE ALL APPROACHED THIS MOST DIFFICULT
OF MISSIONS. WE ARE, INDEED, MOST FORTUNATE IN HAVING SUCH A TEAM
OPERATING ON OUR BEHALF IN BIHAC.

YOURS HAS BEEN A THANKLESS TASK, PERFORMED BRAVELY AND PATIENTLY
WHILE WE IN THE HEADQUARTERS IN SARAJEVO FEEL RATHER HELPLESS,
BUT ONLY DO WHAT WE CAN TO HELP RELIEVE YOUR CIRCUMSTANCES.
PLEASE REST ASSURED THAT WE ARE DOING ALL THAT WE POSSIBLY CAN
IN OUR EFFORTS ON YOUR BEHALF, AND I AM CONFIDENT THAT, AT THE
END OF THIS PARTICULAR DARK PATH IS A BRIGHT FIELD OF OPPORTUNITY
FOR THOSE WHOM YOU SO SELFLESSLY PROTECT.

YOUR OFFICERS AND SOLDIERS HAVE BEEN IN THE FOREFRONT OF THE
WORLDS PRESS AND OPINION IN THE LAST FEW WEEKS. THEY
HAVE BEEN AN EXAMPLE TO US ALL, WHATEVER TACTLESS AND MALINFORMED
INDIVIDUAL REPORTERS MIGHT SAY. I HOPE TO BE ABLE TO VISIT YOU
SOON. I SO ENJOYED OUR LAST MEETING IN SARAJEVO.

Yours most sincerely,

ooi / 062
2225 / A   *Graham Illingworth*

**Personal Memo from a UNPROFOR staff, 1995 (Source: Author's
own wartime records)**

UNPROFOR
UNITED NATIONS PROTECTION FORCE

Lieutenant General Bertrand de LAPRESLE
Force Commander

28 February 1995

Dear General,

The message in your letter to me upon relinquishing the command of the United Nations Protection Force has touched me deeply.

It was a great honour for me to have commanded this most complex mission in the history of the United Nations. I was indeed very fortunate to have been able to count on the assistance and cooperation from all my commanders and their contingents, the staff of both the civilian and military components of the Force, the United Nations agencies as well as other international organizations operating in the Balkans.

I sincerely hope that our efforts will be somewhat instrumental in restoring peace in the former Yugoslavia.

Thank very much for your support.

Let me please, add my deepest appreciation for the outstanding job performed by your very professional and efficient battalion, acting in the very delicate and sensitive area of BIHAC pocket. I have most appreciated the courage and dedication of all : contingent commander, battalion commanding officer and leaders, NCOs and soldiers.

Please believe me very gratefully and faithfully yours.

Lieutenant General Abu Saleh Mohammad Nasim, BB, psc
Chief of Army Staff
Army Headquarters
Dhaka Cantonment

Appreciation letter from Gen. Laprelle

MAGNUS BJARNASON

UNITED NATIONS

UNPROFOR

CIVIL AFFAIRS, BIHAC AREA COMMAND

6 March 1995

Dear Colonel Salim,

Upon my departure from the Bihac Area Command, I would like to take this opportunity to thank you for the time we spent together. I must say that I have been most impressed by your knowledge of the current problems the Bihac Area is plagued by. Your tactful diplomacy has been exemplary in all ways.

I would also like to express my sincere thanks to all your men for their friendliness and willingness to help, often under very difficult circumstances. Especially would I like to thank Major Mahfuz for all his assistance in arranging for accommodation and food, and all the men of the Battalion's Communications Center, who both day and night have been connecting telephone calls and sending messages for Civil Affairs.

I would like to wish you all success in the future and I hope that we may meet again.

Yours sincerely,

Colonel Salim Akhter
Commanding Officer
Bangladesh Battalion
United Nations Protection Force in Former Yugoslavia

Appreciation letter from UN Civil Affairs

NATIONS UNIES
HAUT COMMISSARIAT
POUR LES RÉFUGIÉS

UNITED NATIONS
HIGH COMMISSIONER
FOR REFUGEES

Télégrammes : HICOMREF
Télex : 415740 UNHCR CH
Téléphone : 739 81 11
Téléfax : 731 95 46

Case postale 2500
CH-1211 Genève 2 Dépôt

Colonel Salim Akhtar psc EB,
Commanding Officer,
Banbat.

10 November 1995

Dear Sir,

This is to thank you for all the assistance and cooperation you
have provided to UNHCR during your stay in the Bihac region. This assistance
has been invaluable in executing our programme especially for the delivery of
Humanitarian Aid and supply of fuel to assist to the suppliers of services to
the community.

The conditions under which you operated were often difficult and dangerous
but in spite of these problems you managed to complete the programme for
which we and the beneficiaries are most grateful.

Your role in cooperation with UNHCR has been crucial for the success of our
programme and we request you pass on our special thanks to those officers
and staff who were directly involved with the implementation of our work.

We would like to take this opportunity to wish you a pleasant trip back to
Bangladesh and all the very best for the future of your battalion and for
your families.

Yours sincerely,

Catherine Walker
Deputy Chief of Mission Bosnia.

Appreciation letter from UNHCR

10. Narrative Assessment of Performance:

Colonel Salim has during difficult time in Bihac area performed his tasks and duties clearly above average. The leadership of the unit, which has been exposed to a very different culture and task in FRY, has been based upon modern and strong leadership - taken into full consideration the surroundings and the needed flexibility of BANBAT. The responsibilities and openness toward new tasks have been exemplary and is one of the main factors for the all out good co-operation with local authorities and all UN agencies. Colonel Salim has made a valiant effort to extract the best from a sorry lot.
Colonel Salim is a quiet and thoughtfull commander which advice has been of great help; he has also been the link between external authorities and the Btn. Thereby the colonel was able to be in full control and to direct individuals and units.
It is my firm stand, that the colonel with the amount of experience and with his human approach and openness will be very suitable for future UN postings.

Date: 3/8 - 95

REPORTING OFFICER

Name: H.J.Helsoe

Rank: Colonel

Appointment: COMD/Bihac Area

Signature:

A-2

UN RESTRICTED

UN Annual Report extract, page 1. Report upgraded by General Rupert Smith, Commander, UNPROFOR, , 1995 (Source: Author's own wartime records)

PART IV

11. Comments of Officer being Assessed:

I have been appraised of the contents of my UN Confidential Report. I wish to add the following points.

Signature: _____
Name: SALIM AKHTAR _____
COLONEL
Date: 20 Aug 95 _____
Rank: _____

PART V

12. Comments of Reviewing Officer: I have upgraded this report to reflect my high opinion of Colonel Salim who has brought his unit through a long and difficult tour. He has done this calmly and sensibly showing considerable perseverance and professional competence. When he has stood-in as the Comd of Bihac Area I have always been confident that matters were in a safe pair of hands'.

Signature: Rupert Smith.
Name: Lt Gen R.A. Smith
Rank: _____
    Comd UNPROFOR.
Appointment: _____

A-3

UN Annual Report extract, page 2. Report upgraded by General Rupert Smith, Commander, UNPROFOR, 1995 (Source: Author's own wartime records)

Award of UN Medal

CERTIFICATION

THIS IS TO CERTIFY THAT

COL SALIM AKHTAR, psc

HAS BEEN AWARDED THE UN MEDAL FOR SERVICE WITH UNITED NATIONS PROTECTION FORCE (UNPROFOR) DURING THE PERIOD

The UNPROFOR medal consists of a medallion and a ribbon. The United Nations medallion is in bronze, bears the emblem of the United Nations and the letters "UN" on the front and the inscription "IN THE SERVICE OF PEACE" on the reverse. The ribbon has a blue background, (representing peace), upon which there is a broad red field in the middle bordered by thin white bands (representing UNPROFOR Commands). Centred to the left is a narrow light green field (representing forests) and centred to the right is a narrow brown field (representing mountains).

Bertrand de Lapresle
LtGen
FORCE COMMANDER

UN Medal Certification

Index/Acknowledgements

1. The author's article was published in the Bangladesh newspaper 'The Daily Star' on 24 May 2017.
2. Author's personal notes as commander of The Bangladesh battalion in Bihac, Bosnia Herzegovina; Sep 1994—Nov 1994.
3. Commander Bangladesh Battalion Personal Records, War Time Pictures and War Diary.
4. Col Liaqat Ali Khan, Comd Log Base VK article published in Army Journal.
5. Maj A. K. M. Towhid ul Islam, Comd Bihac Camp, book published in Bangladesh.

www.ingramcontent.com/pod-product-compliance
Lightning Source LLC
Chambersburg PA
CBHW051137120626
46547CB00012B/835